Case Studies in Sport Science and Medicine

Edited by:

Andrew M. Lane, Richard J. Godfrey,
Mike Loosemore, and Gregory P. Whyte

June 2014

ISBN-13: 978-1499146943
ISBN-10: 1499146949

Lane, A. M., Godfrey, R. J., Loosemore, M., & Whyte, G. P (2014). Case Studies in Sport Science and Medicine. ISBN-13: 978-1499146943, ISBN-10: 1499146949. CreateSpace.

Cover image Copyright © of Greg Whyte.

Self-published. Printed by Createspace.

For information regarding the book please contact Prof Andy Lane; A.M.Lane2@wlv.ac.uk

Acknowledgements

The four editors have an accumulated experience of over 100 years in applied sport science and medicine and have been discussing the idea of producing a book that highlights the importance of an individualized, case-by-case approach to optimizing health and performance of athletes and been doing so for some time. The plan was to gain insight into best practice and provide information on elite athlete support from those with first-hand experience and so we contacted the world's leading practitioners and researchers in sport science and medicine to see if they would like to contribute a chapter or comment on a chapter. The production of this book has come about by the fabulous people who have shared their ideas and work at the coal-face of elite sport. The book has been several years brewing and has presented substantial challenges in bringing it to fruition. It is therefore, extremely pleasing to be able to thank all the contributors in public for their unbending support.

Producing a book of so many chapters with contributors from across the globe, from Asia to America and from Western Europe to Eastern Europe, is no small task.

The editors would like to thank Jake Lane for his fantastic work in helping put the book together.

We hope you enjoy the book.

Richard Godfrey, Andy Lane, Mike Loosemore and Greg Whyte.

About the editors

Dr Richard Godfrey

Richard Godfrey worked for 12 years as a physiologist at the British Olympic Medical Centre (BOMC), the last 7 years as Chief Physiologist. He was involved in organising physiology service provision to elite sport before the home-country sports institutes, including the English Institute of Sport, were operational. He has staffed over 130 elite athlete training camps, worked with around 28 different sports in the lab and the field and provided support at many competitions, including numerous World Championships and three Olympic Games. Richard's main research interests include physiology of sports performance, growth hormone and cardiac health. He completed his PhD on 'The exercise-induced growth hormone response in humans and its association with lactate', one study of which was unique in using eight Olympic Gold Medallists as the study population. He is a fellow of ACSM and BASES, has more than 140 peer-reviewed and lay publications and has appeared in the press, and on British and overseas radio and TV, on more than 50 occasions. In 2003, he joined Brunel University where he continues his research and is a Senior Lecturer in Sports Coaching and Human Performance.

Prof Andy Lane

Andy Lane is a Professor of Sport Psychology at the University of Wolverhampton. He is a fellow of the British Association of Sport and Exercise Sciences (BASES), and Chartered Sport and Exercise Scientist. He is Health Professional Council registered and a British Psychological Society Chartered Psychologist. He began lecturing at Brunel University before moving to the University of Wolverhampton in 2000. He has authored more than 200 peer refereed journal articles, edited three books and is a regular contributor to the print, radio and TV media. He has led high profile research projects such as "Can you compete under pressure?" a BBc Lab Uk led project fronted by former Olympian Michael Johnson. His applied work has involved a number of clients, including ranging from recreational to world championship level. He is a member of the UKactive Research Institute's Scientific Advisory Board and has provided consultancy at the Centre for Health and Human Performance at 76 Harley Street, London.

Dr Mike Loosemore

MBBS DCH MRCGP MSc FFSEM(UK) PGCME

Consultant in Sport and Exercise Medicine

Dr Mike Loosemore is a substantive consultant in Sport and Exercise Medicine within the National Health Service (NHS), at the Institute of Sport, Exercise and Health, University College London Hospitals (UCLH).

He is a lead Sports Physician for the English Institute of Sport, where he treats elite athletes from a wide range of sports, and is doctor to the British Boxing. He is also a Sports Physician for Centre for Health and Human Performance at 76 Harley Street, London.

Mike has accompanied teams to Olympic, Commonwealth Games, World and European Championships. He was the Chief Medical Officer for the England Commonwealth Games team in New Delhi 2010 and will again be CMO in Glasgow 2014.

Prof Greg Whyte

In 2014 Professor Greg Whyte was awarded an OBE for his services to Sport, Sport Science & Charity, and was voted as one of the Top 10 Science Communicators in the UK by the British Science Council. Greg is an Olympian in modern pentathlon, and is a European and World Championship medallist. Graduating from Brunel University, he furthered his studies with an MSc in human performance in the USA and completed his PhD at St. Georges Hospital Medical School, London, and the University of Wolverhampton. Greg has published over 250 peer reviewed papers, 8 books and is a regular contributor to the print, radio and TV media. Greg's former roles include Director of Research for the British Olympic Association where he worked with Team GB at 5 Olympic Games, and Director of Science & Research for the English Institute of Sport. Greg is currently a Professor of Applied Sport and Exercise Science at Liverpool John Moore's University and Director of Performance at the Centre for Health and Human Performance at 76 Harley Street, London. He is also the Science Advisor to the Commonwealth Games Committee for England and British Rowing, and the Dean of the Institute of Sport and Exercise Medicine and Chair of UKActive Scientific Advisory Board.

INTRODUCTION

A psychologist meets with an endurance athlete who reports that she is finding training hard, feels she gets tired too quickly and feels she has to stop. The athlete describes a training session where this occurred and it's clear that she is sensitive to physical changes in her body that occur when exercising. When she starts exercising, a warning system that comes in the form of a voice in the back of her mind giving a message of "*hold on, we need to be careful here....hold on........this seems demanding, I need to slow down*"...

What should the psychologist do? Start working on trying to control that inner voice? That is certainly an option. However, the client gives more information on her recent medical history, her training, and her diet. It emerges that she has had an injury that influenced her performance over the previous 12 months, and whilst she thinks it has cleared up, she also acknowledges that she wished she sought help earlier. As the client describes her training programme, diet, and sleep patterns and alarm bells start to ring in the psychologist's mind. It's clear that this is an issue routed in a number of different disciplines. At this point, the psychologist pauses to reflect and consider how to proceed. The psychologist ponders over questions such as; "I wonder what help would be offered if she saw a physiologist and they discussed her training, or what a nutritionist might say about her, or what a physician might say about her injury".

How fellow professionals approached this case would be interesting. In terms of practice, it is common for an athlete to see someone from one discipline and then go onto to see a practitioner from a different discipline. The psychologist could ask themselves "if she went to a physiologist first, how might that affect what I would do, and how I might work?" Practitioners working in sport and exercise science and medicine encounter cases like this frequently. Whilst research provides evidence based upon the mean response of a group of athletes to a given intervention; an individualised approach for a population of N=1 is required to optimise outcome. Each case will be different and require a solution that fits the assessment. How might someone learn the skills and knowledge to do such a task? This book, *Case studies in Applied Sport Science and Medicine* was developed to help fulfil this need.

There has been a need for a book such as *Case studies in Applied Sport Science and Medicine* for some time. Applied Sport Science and Medicine has become an integral part of the high-performance environment in recent years despite the relative youth of the discipline. Early work in sport science and sports medicine attempted to adopt approaches employed in the established disciplines of science and medicine. There has been growing recognition that a multidisciplinary support team should act in an interdisciplinary manner to optimise their impact on health and performance. Athletes require information to be given in a user-friendly way and will tend to see issues in a holistic way, raising questions such as "what does this mean to me?" and "is my training going to produce the results when I want it to?"

Although the terms multidisciplinary and interdisciplinary have become part of normal parlance in the support environment, very few teams provide a truly interdisciplinary approach. This is not criticism of professionals or an excessively bold claim, but recognition of the fact that sport and exercise science and medicine has placed emphasis in its research and practice on developing subject-specific experts, such as sport psychologists, physiologists, and so on. Indeed, the perceived complexity of interdisciplinary support has resulted in many practitioners simply not knowing how an interdisciplinary approach should operate, or even could operate.

Central to the development of a high-quality sport science and medicine support service is the shared understanding of the role of practitioners from different fields. It is not enough to be an expert in your own field; you must be knowledgeable about all aspects of sport science and medicine support and understand how best to utilise the expertise of others to optimise performance. The transferability of knowledge across disciplines is the key to success. Much like the relationship between coach and support staff, the transfer of knowledge is not a one-way process. The ability to explain your discipline-specific work to others is as important as their ability to articulate their intervention. Practitioners, coaches and athletes need to move closer together as a unit in order to enhance knowledge and practice, and develop as a team. Few resources exist to support this transferability of knowledge across disciplines; *Case Studies in Sport Science and Medicine* provides examples of how world-leading practitioners accomplish this complex aspect of support.

We divided the book into three sections.
1. The Reactive Model: Providing solutions for pre-existing problems
2. The Predictive Model: Providing solutions for events that are predicted to occur
3. The Proactive Model: Providing on-going support and developing interdisciplinary teams

In the first section we look at how practitioners work with pre-existing issues. It is common for practitioners to work with clients with issues that have been diagnosed and identified. In many ways, this is the "textbook" approach as it allows the practitioner to do some research on the issue, find out the latest research and then consider whether to use in her or his practice. However, the nature of applied work means that each case is unique and the application of one treatment to what can initially look like a similar condition can throw up a plethora of unknowns and a number of "ifs and "buts." The take-home message is that the treatment presented by the practitioner is very different to what it looks like in the manual; or it can be. This section of the book illustrates this issue and brings some world-leading practitioners to the table.

The second section focuses on a number of themes in which the future competitive environment is known and so can be planned for. These include preparing to cope with the rigours of extreme climates, preparing for success in multiple events at one competition, addressing training either to boost performance or prevent injury and addressing the challenge of optimal hydration and nutrition. Without appropriate planning for "any and all" eventualities, preparation is suboptimal and there are greater risks then that the athlete will not perform to his or her best. The World's best athletes and their coaches in any given sport have two major things in common: the best athletes are all very similar in terms of anatomy and physiology and the best coaches are excellent organisers, managers and strategists. So in preparing for anything at the elite level (competition or training) 'leave no stone unturned' is the all-important mantra. To achieve this, the requirement for expertise has grown as there is greater need for more and more accurate assessment and interpretation of the specialised sports science and medicine data that increasingly contributes to diminishing margins of success.

In the third section, we examine how sports scientists provide ongoing support. For example, a sports medic is working with the client, possibly listening to her or him describe the rehabilitation training from an injured knee. The consultant gets a hint that the client is struggling with the injury, and that rehabilitation is not being done at the intensity that it should be. The athlete appears to be saying what is expected rather what has happened. The sports medic wishes he had a sports science team sitting in residence in his mind, so she/he could address the issue from an interdisciplinary perspective. This section of the book works along those lines, detailing cases where practitioners are working in an interdisciplinary way and as such offers some fabulous insights into applied work.

In recognition of the need for a bespoke, individual tailored approach, this book examines examples of support from a case study perspective across the broad range of sport science and medicine disciplines written by recognised world leaders. This book provides 29 case studies covering physiology, psychology, biomechanics, motor control and performance analysis, nutrition, strength and conditioning and sports medicine. Each case study is presented in a structured format providing a vignette of the case with key information including the challenges faced. The vignette is followed by a contemporary review of the key literature in the field informing the decision-making process involved in the case study and related differential diagnoses and interventions. The case study is concluded by presenting the intervention and outcome. Each case study is followed by a commentary from another world leader drawing out salient points, expanding the discussion and giving personal insight.

Practitioners, athletes, students and anyone interested in sport should find the content of these case studies relevant and useful; they are diverse and capture the range of issues consultants face. Overall, *Case Studies in Sport Science and Medicine* offers a unique and valuable collection of case studies in a wide range of sport science and medicine disciplines written by world leaders in the field of high-performance sport for those working in the field of sports science and medicine.

CONTENTS

Part III: The Proactive Model: Providing on-going support and developing Interdisciplinary teams

119

Part I: The Reactive Model: Providing solutions for pre-existing problems

CHAPTER 1

Sick and Tired of Being Sick and Tired: Case Study of an International Kayaker's Recovery from Chronic Fatigue Syndrome and Psychological Preparation for the World Championship

Peter Terry
Centre for Health Sciences Research, University of Southern Queensland, Toowoomba, Australia.

Vignette

Psychological support was provided over a 15-year period spanning 1994-2008 to a Caucasian female, 6-time world champion and marathon kayaker. Throughout much of this period, she presented with symptoms associated with overtraining and unexplained underperformance (1), which culminated in a clinical diagnosis of chronic fatigue syndrome (CFS) in 2003 when she was 26 years old. Difficulties associated with diagnosis and treatment of CFS are well documented (2). The etiology of CFS remains inconclusive, laboratory markers are unreliable and an effective cure is elusive (3). Broader conceptualisations of CFS that view it as a biopsychosocial condition (4) or an adjustment disorder (5) have challenged the traditional conceptualisation of CFS as a biological imbalance caused by excessive training and/or inadequate recovery. Following two years of ineffective treatment, during which time one sports medicine specialist told her she may never race again and that her body was saying "it's time to hang up your paddles", the athlete eventually recovered in 2005 having followed a reverse therapy regimen (6). Reverse therapy treats CFS as a disorder of the hypothalamus-pituitary-adrenal axis and provides an educative process focusing on such things as the link between emotions and health, finding a balance between training load and life demands, and emphasising fun and lightness in her approach to training.

Mood disturbance is acknowledged as a common symptom in overuse conditions (7). Hence, mood profiling has been advocated as a useful monitoring tool for athletes generally (8) and for elite kayakers in particular (9). Mood scores for fatigue, vigour and depression have been shown to be particularly germane in this context (9,10). During the period of rehabilitation from CFS and in preparation for the 2005 world championship, music was used to regulate effort levels, control attentional focus, and generate specific emotional responses (11). Goal-setting exercises (12) were used to establish milestones for recovery, to provide a framework for her preparation for the world championship, and to clarify a specific race plan. Relaxation and guided imagery exercises (12) were used to mentally rehearse her race plan for the world championship, in which she made a successful return to competition at the highest level. These techniques were also applied to helping her manage non-sport stressors with a view to maintaining a robust sense of self-esteem that was founded on her qualities as a person and her life beyond sport.

Discussion

This case raises issues concerning (a) the almost inevitable tension between performance demands and athlete health, (b) coaching practices and athlete well-being, and (c) medical and psychological support for athletes. Her coach was a very experienced, highly successful practitioner who, while sensitive to her vulnerability to overuse conditions, based his program on high training volume even by the demanding standards of marathon kayaking. In hindsight, the training demands made in the name of performance enhancement may not have been conducive to her physical health and psychological well-being. This observation should not be interpreted as criticism of the coach, whom I would describe as a hard-driving but caring taskmaster. Instead, it points to the need to include regular input from health professionals who place athlete well-being as their highest priority while recognising the difficult balancing act required of coaches to achieve the maximum training effect for athletes but not burning them out in the process.

During the two-year period that she spent fighting the debilitating effects of CFS, telephone counselling was provided on a regular basis. Given her persistent fatigue and susceptibility to recurrent infections, her emotional state for much of this period was one of confusion with moments of despair, although no clinical depression was diagnosed. Although the inter-individual variability of a spectrum of symptoms for CFS is acknowledged (1), concomitant individualisation of treatment plans does not necessarily occur. Frustration at prolonged lack of progress via conventional medicine caused her to seek alternative treatment, including acupuncture, hypnotherapy, crystal therapy and spiritual healing. After suffering, as she described it, "15 months of terror and error", she embarked on a course of reverse therapy (6) which helped her overcome the illness.

In an account of her struggle with CFS (13), she recalls how the same personal strengths that helped her become world champion – commitment, single-mindedness, focus, dedication – had contributed to the demise of her health. Her vigilance, bordering on obsession, in always doing the right thing in training, diet, sleep patterns and so on, drained her happiness and love of kayaking to the point where she felt permanently exhausted. During her lowest moments, lying on her bed in tears, she gained much from mental imagery. "I'd visualise a race course, see myself on the start line, picture my rivals and see the race unfold. I'd see everything; I'd imagine the weather, but I wouldn't just see it. If it was raining, I would feel the rain on my skin and feel the wind in my face. I used all my senses. And of course I'd always win ... It was very powerful; not only did it inspire me and cheer me up but also I was creating movies in my mind that I would use when I was racing again".

In the 2005 world championship, held in Perth, Australia, women completed 4 laps of a 7.2 kms course, punctuated by 4 portages where they ran 200 m overland carrying their kayaks before re-entering the water to continue paddling. The athlete was renowned for her fast portages and accordingly she based her race plan on stay quietly on pace with the leading pack, running very hard at the final portage and then burning (making a sustained effort) over the next 4 minutes. Plan A was that she would drop all her competitors at that stage, leaving them to fight for the minor placings. Plan B was to beat her rivals in a sprint finish as she had done previously in the European championship. Plan A relied on feeling good coming to the final portage whereas Plan B was seen as higher risk due to the relatively unknown sprinting capacity of some of her rivals. Commitment to Plan A grew in the 48 hours preceding the race, reinforced by regular mental rehearsal of her performance of the decisive portage.

Using rational-emotive behaviour therapy (14) incorporating a solution-focused approach (15), many of her anxieties about the forthcoming race were challenged. These included doubts about her physical condition, her ability to reproduce past form and, in particular, whether she would "seize up" during the race as she had experienced several times in training. During frequent discussions about her race plan, which had already been endorsed by her coach, several messages of confidence were drip fed, including that she was the most experienced competitor, the best tactician, the one they all feared, the one with the record of winning, the one who wanted it the most, and that no one could drop her and no one was more determined.

The athlete's mood responses were monitored using the Brunel Mood Scale (16). Figure 1 shows her mood profiles leading into the world championship. Profiles demonstrate increasing vigour and decreasing fatigue; which from her previous mood profiles were known to be positive indicators for this athlete. Depressed mood and anger remained stable at minimum levels. Confusion was eliminated as race day came around due largely to her race plan becoming more clearly understood and well rehearsed. Tension fluctuated but remained within the normal range for athletes, considering the importance of the event. Overall, her pre-race mood profile was consistent with theoretical predictions of excellent performance (8) and with optimal mood profiles associated with her previous successful performances.

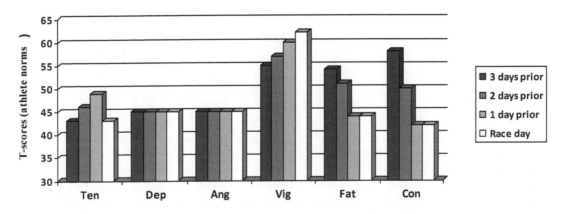

Figure 1. Mood profiles demonstrating mood enhancement leading into the world championship

As for music interventions, during her period of rehabilitation from CFS and in preparation for the world championship, the athlete used different songs for different purposes. While training on-water wearing sunglasses with a built-in mp3 player, she used Nelly Furtado's *I'm Like a Bird* to (a) regulate her stroke rate by matching it to the tempo of the song, (b) generate feelings of being relaxed and free, and (c) reinforce the metaphor of learning to "fly" again following her debilitating condition. She used songs such as *Happy Face* (Destiny's Child) and *Rise Up* (Yves Larock) to feel uplifted and happy, and the theme from the *Rocky* movies when motivation began to wane. R. Kelly's *World's Greatest*, the theme song from the movie *Ali*, was reserved for race day. In her words, "I absolutely love the lyrics. I save this song for the really big races. I only ever listen to it at the world championships when I am warming up for my race. It brings a lump to my throat; it puts me in the zone"

Conclusions

In this case, conventional medical approaches were ineffective in treating CFS whereas an alternative approach, reverse therapy, proved efficacious. Although reverse therapy is based on established neuroscientific principles, the evidence base for its efficacy is not well established empirically. Further evaluative research is therefore warranted. Feedback from the athlete supported use of a range of sport psychology strategies, including mood profiling, goal-setting, imagery, and music interventions. Meta-analytic summaries of the research literature (17-20) have strengthened the evidence base for such techniques. The weighted mean effect sizes in these meta-analyses indicate robust benefits to performance and/or psychological well-being in the small-to-moderate range for goal-setting (17), music interventions (18) and mood profiling (19), and in the moderate range for imagery (20). The present case study has highlighted the benefits of a multidisciplinary approach to supporting athletes' health needs and emphasises the potential efficacy of alternative medical approaches.

References

1. Budgett R, Newsholme E, Lehmann M, Sharp C, Jones D, Jones T, Peto T, Collins D, Nerurkar R, White P. Redefining the overtraining syndrome as the unexplained underperformance syndrome. *Brit J Sports Med* 2000;**34**:67-68.
2. Shephard RJ. Chronic fatigue syndrome: A brief review of functional disturbances and potential therapy. *J Sport Med & Phys Fit* 2005;**45**:381-392.
3. Wallman KE, Morton AR, Goodman C, Grove R. Reliability of physiological, psychological, and cognitive variables in chronic fatigue syndrome. *Res Sports Med* 2005;**13**:231-241
4. Schwenk TL. The stigmatization and denial of mental illness in athletes. *Brit J of Sports Med* 2000;**34**:4-5.
5. Jones CM, Tenenbaum G. Adjustment Disorder: a new way of conceptualizing the overtraining syndrome. *Int Rev Sport & Exerc Psychol* 2009;**2**:181-197.
6. Eaton J ME, *Chronic fatigue syndrome and fibromyalgia: The reverse therapy approach*. London, Authors Online; 2005.
7. Morgan WP, Costill DC, Flynn MG, Raglin DS, O'Connor PJ. Mood disturbance following increased training in swimmers. *Med Sci Sports Exerc* 1988;**20**:408-414.
8. Terry PC, Lane AM. Research and application in the area of mood and emotions in sport and exercise. In: Morris T, Terry PC, eds. *Sport and exercise psychology: The cutting edge*. Morgantown, WV: Fitness Information Technology; 2010:204-229.
9. Kenttä G, Hassmén P, Raglin RS. Mood state monitoring of training and recovery in elite kayakers. *Eur J Sport Sci* 2006;**6**:245-253.
10. Lane AM, Terry PC. The nature of mood: Development of a conceptual model with a focus on depression. *J Appl Sport Psychol* 2000;**12**:16-33.
11. Karageorghis CI, Terry PC. The psychological, psychophysical, and ergogenic effects of music in sport: A review and synthesis. In Bateman AJ, Bale JR, eds. *Sporting sounds: Relationships between sport and music* London: Routledge; 2009:13-36.
12. Karageorghis CI, Terry PC. *Head strong: The art and science of applied sport psychology*. Champaign, IL: Human Kinetics; 2010.
13. Barton A. *Recovery from CFS: 50 personal stories*. Milton Keynes, UK: Author House; 2008.
14. Ellis A. *Overcoming resistance: A rational emotive behaviour therapy integrated approach*. New York: Springer; 2007.

15. Guterman JT, Rudes J. A solution-focused approach to rational-emotive behavior therapy: Toward a theoretical integration. *J Rational-Emotive & Cog Behav Therapy* 2005;**23**:223-244.
16. Terry PC, Lane AM, Fogarty GJ. Construct validity of the POMS-A for use with adults. *Psychol Sport & Exerc* 2003;**4**:125-139.
17. Kyllo LB, Landers DM. Goal setting in sport and exercise: A research synthesis to resolve the controversy. *J Sport & Exerc Psychol* 1995;**17**:117-137.
18. Terry PC, Lim J, Mecozzi A, Karageorghis, CI. Meta-analysis of effects of music in sport and exercise. In *Proceedings of the ISSP World Congress of Sport Psychology*. Marrakech, Morocco: International Society of Sport Psychology; 2009;157.
19. Beedie CJ, Terry PC, Lane AM. The Profile of Mood States and athletic performance: Two meta-analyses. *J Appl Sport Psychol* 2000;**12**:49-68.
20. Curran ML, Terry PC. What you see is what you get: A meta-analytic review of the effects of imagery in sport and exercise domains. In *Proceedings of the International Congress of Applied Psychology*. Melbourne, Australia: International Association of Applied Psychology; 2010.

Commentary: *Chris Beedie*, University of Aberystwyth, UK

Peter Terry's case study describes one of a group of sports-related illnesses, including overtraining syndrome and burnout, which whilst historically conceived as physiological conditions, likely have as much in common with mental health syndromes, specifically mood disorders. In many respects, the situation Peter describes, that of an endurance athlete who is trapped in a cycle of increasing efforts and diminishing returns, is not unusual. However, in this case, the athlete herself suggests that the very traits and behaviours she associated with elite performance, for example, dedication, focus and commitment, contributed to a substantial worsening of her health.

The mentality of elite performers is often almost dualistic, one of the self and the body as separate entities, with the self being able to drive the body towards ever better functioning, the brain essentially serving to program one's body. The reality of course is that the 'self' is a function of mind, which is itself a function of the brain. To function properly, the brain is dependent on the same physiological processes as are the heart, and the musculo-skeletal and cardio-respiratory systems. Therefore, when physiological processes break down, the self is as at risk as the body.

Peter Terry, recognising the totality of the condition, as well as the potential impact of a broad range of factors in its genesis, employs a long-term, multi-modal and multi-phase approach to treatment. He incorporates interventions from traditional cognitive-behavioural approaches such as goal setting and imagery, through to the use of music as both a symbolic and emotive self-regulation mechanism during training and prior to performance. Music in this respect is not a proxy; music is part of the therapy. Significantly, Peter's approach is one in which the psychological skills developed by the athlete during the process were as much applicable to the athlete's life beyond sport as to the immediate competitive environment. No doubt these skills were likely essential factors in both her journey through illness and the return to the status quo, free of these problems.

Peter Terry suggests that his case study raises issues concerning "the almost inevitable tension between performance demands and athlete health", suggesting a potential clash between performance and well being interests. The physical and mental health of an elite athlete in a period of intense or important competition often rests on a knife edge; a move away from the edge in one direction might represent improved physical and mental health and an oasis of calm, but below-par sports performance, and an athlete (and coach) would likely be motivated to avoid such a scenario. However, a slight move from the edge in the other direction might represent a substantial deterioration in that athlete's ability to function, both within and beyond sport, and the athlete could find themselves in dire straits. Within the highly charged sport environment, this type of deterioration is all too often addressed in the context of sports performance, and strategies aimed at reversing such a decline are often counter-productive leading to further declines. As Peter suggests, such a situation calls for the intervention of a health professional, and not a rush by the coach to police the athletes's training load in an attempt to get performance back on track in a blur of sport-related activity.

Peter's account highlights the increasingly evident role of mood measurement in the evaluation of athletes. Peter suggests that mood disturbance is a symptom in overuse conditions. This argument could be extended to stating that mood might in fact be an evolved signal of that developing condition, a process that functions to tell us, and others, that our resources are low. Interestingly, the mood measures Peter used, often used as predictors of performance in sports research, were initially developed as a measure of mood disorder in clinical psychiatry.

The case study described by Peter Terry leaves one in no doubt that, ultimately, his focus on a range of broad issues including sport, was ultimately a more effective sports psychology intervention than would have been the case had he adopted a narrow focus on performance issues alone. It highlights the need for sports practitioners to understand the complex interaction of biological, psychological and sociological processes in maintaining both elite performance and the optimal mental and physical health of athletes, or in suggesting the cure for any such condition as that described.

CHAPTER 2

The Use of Strength and Conditioning in Reducing Shoulder Pain and Improving Function in a Wheelchair Athlete

Paul Gamble
Sports Performance Research Institute New Zealand, AUT University, Auckland, New Zealand.

Vignette

This case study describes a 58-year-old male Paralympic wheelchair athlete (part of the GB curling squad). He is paraplegic with an incomplete spinal lesion at T12/L1 as the result of a car accident when he was a teenager. The initial medical briefing described the athlete as possessing some lower-limb function but nevertheless dependent on the wheelchair, albeit able to use crutches for a short period of time. The athlete has complete upper limb function was also described as having 'some degree of core control/stability, enabling him to balance in sitting at a reasonable level'.

Discussion

The athlete was undergoing physiotherapy treatment having presented with left (non-dominant) shoulder pain. This was an on-going complaint: the athlete reported the symptoms had been gradually increasing over a four-year period; but had become markedly worse over the previous four months. The athlete described a pain that disturbed his sleep through the night with a painful restriction to shoulder movement. The clinical assessment reported focal tenderness of the superior, posterior aspect of the acromioclavicular (AC) joint with associated painful restriction of shoulder ROM in flexion, external rotation, and abduction. In addition, subacromial impingement to the bursa was evident. Further medical investigation via ultrasound revealed degenerative changes to the AC joint.

Following these investigations, an orthopaedic referral advised surgical intervention. However, the athlete's circumstances made this treatment option highly problematic. Both the athlete and his wife were wheelchair bound and so providing care in the post-operative period would be very difficult without external assistance. The original clinical assessment highlighted areas for improvement in terms of the athlete's seating posture in the chair and both the resting position and dynamic control of the shoulder girdle:

a. Postural assessment: 'Protraction of the cervical spine, increased thoracic kyphosis, slumped seating position in wheelchair (with back rest slack and unsupportive)'.

b. Musculoskeletal assessment: 'Cervical spine restriction of right rotation at 55 degrees and right lateral flexion to 15 degrees. Muscular tightness through scalenes (left), pectoralis minor (left > right), major (left > right) and levator scapulae (left > right) was also noted.'

This suggested that appropriate treatment and corrective exercise – i.e. mobilization and manual therapy with strength/neuromuscular training – to address these issues might improve the athlete's symptoms.

Initial Referral and Intervention Plan

In recognition of the constraints described the lead physiotherapist was understandably very keen to explore non-operative treatment/intervention, at least initially. The approach suggested was consistent with current thinking regarding the management of degenerative musculoskeletal disorders of the shoulder, namely, the use of exercise, in combination with manual therapy (1). To this end, I was approached to work with the athlete in my capacity as a strength and conditioning specialist, with a specific interest in corrective exercise prescription.

Based upon clinical assessment, it was felt there was considerable scope to improve the resting position of the athlete's shoulder girdle, and the strength and neuromuscular control of the posterior rotator cuff. The initial guidance given by the physiotherapist was to take a graduated approach in terms of exposure to movement/loading challenges, and to work within pain-free ranges of motion. In view of the nature of the injury, it was important to avoid internal rotation and work towards end range of motion in shoulder flexion. More specifically those exercises, which involved loading the shoulder overhead, were contraindicated and so were also avoided.

The athlete was introduced to a graduated program of strength training specifically aimed to improve shoulder control, stability, and function. The initial training intervention met with positive results and the decision was made to continue with conservative management of shoulder pain employing mobilisation and exercise in a treatment protocol. In addition to on-going physiotherapy treatment, the initial medical intervention was a cortisol injection, given under ultrasound guidance.

It was also identified that the athlete was carrying excess fat mass, and was essentially overweight. There is an established link between incidence of shoulder pain and increased body mass. Higher body mass is associated with greater loading of the shoulder joint during the pushing phase of propulsion and signs of shoulder joint pathology in wheelchair users (2). In accordance with these findings, interventions to reduce body mass were suggested to help reduce the prevalence of pain and injury.

The primary objective of the medical/physiotherapy and corrective exercise intervention was for the athlete to achieve active, pain free, range of motion and ultimately commence the competition season with minimal shoulder pain and hence with no sleep disturbance. Therefore, the final decision regarding surgical intervention would be made following clinical assessment and a review of progress at a specified date, based upon these criteria.

Strength Training Intervention

The programme consisted of a range of exercises performed when seated (i.e. with the athlete in his own wheelchair). Training modes consisted of cable resisted and free weight (dumbbell) exercises, avoiding pressing or resisted raising movements performed overhead. Specifically, the programme featured two main elements:

1. Pulling movements in different planes and pressing activities (avoiding overhead movement), with an emphasis on shoulder girdle positioning and control throughout each repetition;
2. A range of scapula stabilisation and rotator cuff exercises selected from the literature (3), with some modifications.

The training programme comprised a combination of both isolated exercises for specific development of rotator cuff muscles and more complex exercises that were progressively introduced which allowed for greater force development and a greater degree of synergistic muscle action. This approach reflects what has been advocated previously in the literature (4).

Exercises were predominantly conducted in an alternate-arm or single-arm fashion. The athlete was cued to fix the contralateral scapula whilst focussing on position and control of the shoulder girdle and upper limb performing the resisted action. Likewise, the athlete was encouraged to focus on retaining posture and shoulder girdle positioning during the concentric phase with control during the eccentric phase. Loading in this way also imposed an additional postural training stimulus as the athlete was challenged to maintain torsional stability in order to minimise any extraneous movement during the alternate-limb or single-limb action.

Following the start of the strength training intervention the athlete was referred to a performance nutritionist. Likewise, once symptoms began to resolve to the extent the athlete could use an arm ergometer without any shoulder pain, the athlete was provided with supplementary metabolic conditioning sessions. That is the combination of appropriate nutrition and rhythmic aerobic exercise was used in an attempt to reduce body mass, and so reduce the stress on the athlete's shoulders when propelling himself in the wheelchair.

Outcomes

The athlete immediately perceived a benefit from performing the corrective exercise intervention; as a result the athlete's attendance at and compliance with the strength training sessions was excellent.

The follow-up physiotherapy/medical review that occurred six months post intervention was very positive as seen in the quote below:

'*As you should now all be aware TK has opted not to have surgery for the time being. I have spoken with him and he is happy for me to communicate with you all to ensure that his management move forward is as co-ordinated as possible with the main aim to minimise the risk to the shoulder region.*'

The strength training intervention was therefore continued and the programme was progressed accordingly, in terms of load, degree of neuromuscular challenge, and range of motion. The nutritional intervention and conditioning also produced favourable results, with the athlete achieving significant (20kg) weight loss in a seven-month period.

The athlete was able to successfully compete in the Winter Paralympics competition in Vancouver in 2010. To date, the athlete has not been required to undergo surgery, and he continues to train and compete at the highest level, subsequently being selected for the Sochi Winter Paralympic Games in 2014. A graduated increase in load and stabilisation challenge was applied over a four-month period; the major focus remained on posture and shoulder girdle positioning throughout.

Exercises	Cable Set Up	L	R	31-Aug-09			07-Sep-09			14-Sep-09			21-Sep-09		
				W-up	sets	reps	W-up	sets	reps	W-up	sets	reps	W-up	sets	reps
Session 1															
Single Arm Rows	5A	5	6		3	8		3	8		3	8		3	8
Cable Press	2A	4	4		3	8		3	8		3	8		3	8
Alternate Arm Cable Reverse Fly	7B	2	3		3	8		3	8		3	8		3	8
Alternate Arm Dumbbell Full Can Raise		4	5		3	8		3	8		3	8		3	8
(Alternate Arm) Straight Arm Cable Pulldown	10B	3	3		3	8		3	8		3	8		3	8
ONE ARM Dumbbell Bicep Curl		8	8		3	8		3	8		3	8		3	8
Session 2															
Cable Shoulder External Rotation	5A	1	1		3	7		3	7		3	7		3	7
Cable Diagonal Pulley	8B	3	3		3	7		3	7		3	7		3	7
Alternate Arm Cable Reverse Fly	7B	2	3		3	7		3	7		3	7		3	7
Alternate Arm Dumbbell Full Can Raise		4	4		3	7		3	7		3	7		3	7
(Alternate Arm) Straight Arm Cable Pulldown	10B	3	3		3	7		3	7		3	7		3	7
ONE ARM Dumbbell Bicep Curl		8	8		3	7		3	7		3	7		3	7

Exercises	Cable Set Up	L	R	28-Sep-09			05-Oct-09			12-Oct-09		
				W-up	sets	reps	W-up	sets	reps	W-up	sets	reps
Session 1												
Single Arm Rows	5A	5	6		3	8		3	8		3	8
Cable Press	2A	4	4		3	8		3	8		3	8
Alternate Arm Cable Reverse Fly	7B	2	3		3	8		3	8		3	8
Alternate Arm Dumbbell Full Can Raise		4	5		3	8		3	8		3	8
(Alternate Arm) Straight Arm Cable Pulldown	10B	3	3		3	8		3	8		3	8
ONE ARM Dumbbell Bicep Curl		8	8		3	8		3	8		3	8
Session 2												
Cable Shoulder External Rotation	5A	1	1		3	7		3	7		3	7
Cable Diagonal Pulley	8B	3	3		3	7		3	7		3	7
Alternate Arm Cable Reverse Fly	7B	2	3		3	7		3	7		3	7
Alternate Arm Dumbbell Full Can Raise		4	4		3	7		3	7		3	7
(Alternate Arm) Straight Arm Cable Pulldown	10B	3	3		3	7		3	7		3	7
ONE ARM Dumbbell Bicep Curl		8	8		3	7		3	7		3	7

Figure 1. Sample Programme

References

1. Green SR, Buchbinder SE. Physiotherapy Interventions for Shoulder Pain (Review), *The Cochrane Library*, Issue 9, 2010.
2. Collinger JL, Boninger ML, Koontz AM, Price R, Sisto SA, Tolerico MA, Cooper RA. Shoulder Biomechanics During the Push Phase of Wheelchair Propulsion: A Multisite Study of Persons with Paraplegia, *Archives of Physical Medicine and Rehabilitation*, 2008;**89**(4): 667-676.
3. Escamilla RF, Yamashiro K, Paulos L, Andrews JR., Shoulder Muscle Activity and Function in Common Shoulder Rehabilitation Exercises, *Sports Medicine*, 2009;**39**(8): 663-689.
4. Giannakopoulos K, Beneka A, Malliou P, Godolias G. Isolated vs. Complex Exercise in Strengthening the Rotator Cuff Muscle Group, *Journal of Strength and Conditioning Research*, 2004;**18**(1): 144-148.

Commentary: *Stuart Miller*, University of Bath, UK.

Shoulder injuries in wheelchair users are not uncommon (1,2) the effect on sport and lifestyle can, however, be much more significant than a similar injury in an able-bodied athlete. Getting out of bed, using a wheelchair and other activities of daily living can be seriously compromised. Whilst it may be a treatment of first choice, intervention through surgery can significantly compromise the ability of the athlete to live, work and train independently for a prolonged period.

In managing an injury in Paralympic athletes as with any athlete, one must consider the factors that affect optimum functioning of that person and what can be altered to improve that function. Whilst there is a tendency to concentrate on *dis*ability, it is much more important to look at what the individual *can* do and to understand the functional *ability* in relation to the spinal cord injury (SCI) whilst being aware of the effect of loss of motor control and its effect on the wider biomechanical function.

In this type of situation, a number of important factors may play a role:
- The T12/L1 lesion will result in a lack of lower lumbar and pelvic stability. In this situation, it is reported as incomplete – a full understanding of the ability below the spinal injury level is important.
- The strength and stability of the shoulder;
- The alignment of the shoulder;
- The overall biomechanical effectiveness of the athlete, including overall posture;
- The loads applied across the joint including compounding factors such as the athlete weight and the influence on wheelchair propulsion.

Shoulder function requires not only the muscles that move the shoulder to be intact but also the muscles that provide stability around the shoulder girdle to function (supplied by C4-T2 myotomes). In this case, the shoulder musculature is unaffected by the spinal cord injury and can be effectively trained to improve strength and stability and to restore optimal muscle balance whilst encouraging proper gleno-humeral alignment and addressing any tight anterior soft tissue structures.

The shoulder also needs to maintain an optimum joint position to allow full function. The kyphotic posture with cervical protraction as described is not uncommon in SCI, and can lead to potential shoulder impingement due to the position of the glenoid relative to the desired plane of movement. Encouraging a more upright posture through truncal stability training will ultimately help to align the shoulder. Optimizing seat stability and back-rest design both in their normal 'day' chair and any 'competition' chair will help to give an external boost to overall truncal stability (3).

It is also important to look at the ability and normal movement patterns of an athlete during truncal stabilization such as by weight counterbalancing or the use of the contralateral upper limb to counter truncal rotation and flexion. This is sometimes an important factor in developing effective strong and stable sporting movement and may need to be taken into account during stability training.

In conclusion, and to quote a Paralympic athlete I was treating, 'It's all about function' – not disability. If you can understand what an athlete can do, then it will go a long way to restoring their optimum performance.

References

1. Sie IH, Waters RL, Adkins RH, Gellman H. Upper extremity pain in the postrehabilitation spinal cord injured patient. *Arch Phys Med Rehabil.* 1992;**73**(1):44-48.
2. Curtis K, Drysdale G A, Lanza D R, Kolber M, Vitolo R S, West R, Shoulder pain in wheelchair users with tetraplegia and paraplegia, *Arch Phys Med Rehabil,* 1999;**80**(4):453-457.
3. Pavec D, Aubin CE, Aissaoui R, Parent F, Dansereau J. Kinematic modeling for the assessment of wheelchair user's stability. *IEEE Trans Neural Syst Rehabil Eng.* 2001;**9**(4):362-368.

CHAPTER 3

Hypertrophic Cardiomyopathy and Ultra-Endurance Running - Two Incompatible Entities?

Sanjay Sharma[1] and Mathew G. Wilson[2]
[1] St George's University of London, Division of Cardiac & Vascular Sciences, London, UK
[2] ASPETAR, Qatar Orthopaedic and Sports Medicine Hospital, Doha, Qatar.

Vignette

Regular physical exercise is associated with physiological increases in cardiac dimensions, which may be reflected on the electrocardiogram (ECG). Differentiating a physiological or pathological remodelling mechanism is important, as significant cardiac enlargement may be an expression of underlying cardiac disease, placing the athlete at a greater risk of sudden cardiac death (SCD) [1]. Hypertrophic cardiomyopathy (HCM), defined by the presence of increased ventricular wall thickness or mass in the absence of loading conditions (hypertension, valve disease, etc.) sufficient to cause the observed abnormality [2], is the leading cause of SCD in the young and accounts for one-third of all sudden cardiac deaths in young competitive athletes [3, 4]. However, existing data also demonstrates that a small proportion of athletes (<2%) exhibit increased left ventricular wall thickness (LVWT) ranging between 13-16mm [5-7], which overlaps with morphologically mild HCM. Deaths from HCM are predominantly confined to intermittent 'start-stop' sports such as American football, basketball and soccer, with few cases reported in endurance sports. The postulated theory is that individuals with HCM are unable to augment cardiac output sufficiently to participate in intensive and prolonged endurance sports due to a combination of pronounced LVH, a non-compliant LV, exercise-induced LV outflow obstruction and microvascular ischemia. However, we report an ultra-endurance athlete with confirmed HCM, capable of performing high-levels of aerobic ultra-endurance activity.

Case Presentation

A 44-year-old Caucasian male was evaluated in our centre for investigation of a cardiac murmur identified by his primary-care physician. The individual was asymptomatic with no past medical history, medication history or family history. He was an ultra-marathon runner with over 25 years of competitive running history; currently participating in 3 ultra-marathon (>50km) events per year, often involving challenging mountainous and frozen terrain. Resting blood pressure of 95/60 mmHg and physical examination was unremarkable apart from the presence of a soft ejection systolic murmur.

The ECG demonstrated first-degree heart block, right axis deviation, voltage criteria for bi-atrial enlargement, LVH and significant repolarisation anomalies including ST-segment depression in leads II, III and AVF, and deep T-wave inversions in leads V5 and V6 (Figure 1). Echocardiography demonstrated asymmetric septal hypertrophy of the basal and mid-septum with a maximal LVWT of 14mm and an end-diastolic LV diameter of 44mm (Figure 2). There was no evidence of systolic anterior motion of the mitral valve leaflet or LV outflow tract obstruction. Systolic and diastolic function were normal; the left atrial diameter measured 37mm, the E/A ratio was >1 (Figure 3) and tissue Doppler revealed an E' of 16 cm/s at the lateral LV wall and 11 cm/s in the septal LV wall (Figure 4).

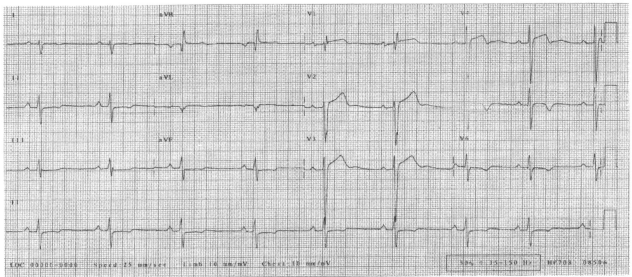

Figure 1: 12-lead ECG of a 44-year-old ultra-marathon runner demonstrating first degree heart block, right axis deviation, bi-atrial enlargement, left ventricular hypertrophy with associated ST-segment depression in leads II, III, AVF and deep T-wave inversions in leads V5 and V6.

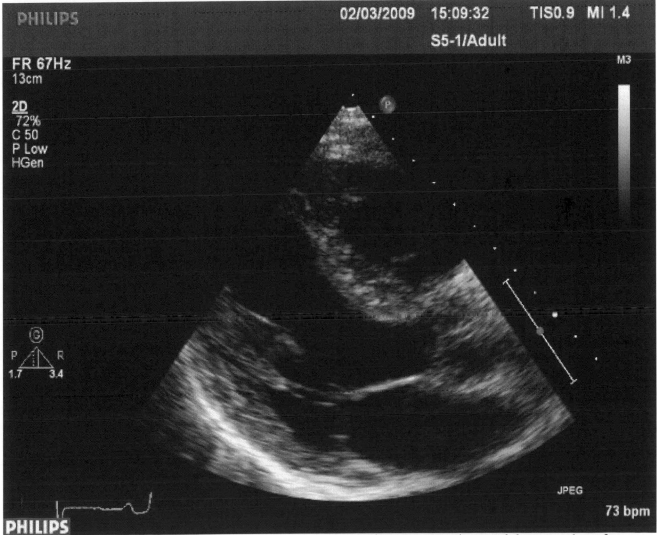

Figure 2: Trans-thoracic echocardiography images demonstrating; asymmetric septal hypertrophy of 14mm and a left ventricular cavity size of 44mm in the parasternal long axis and short axis at papillary muscle level.

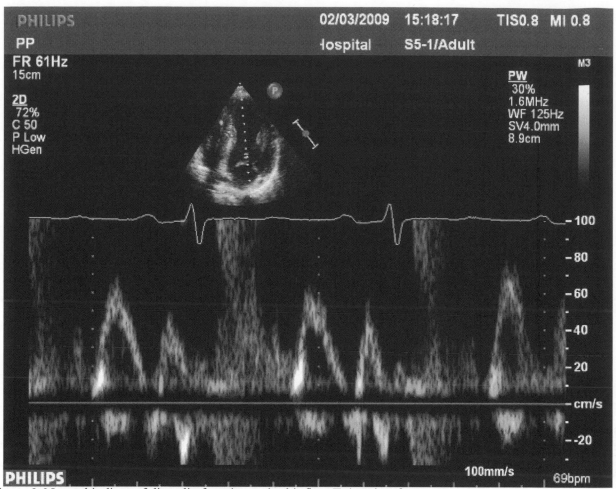

Figure 3: Normal indices of diastolic function; mitral inflow E:A ratio of >1

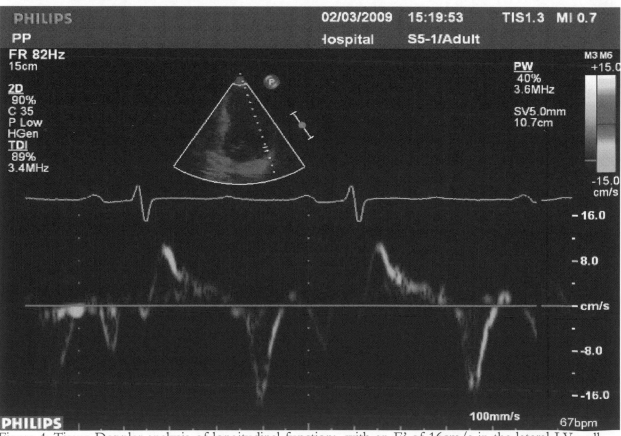

Figure 4: Tissue Doppler analysis of longitudinal function; with an E' of 16cm/s in the lateral LV wall and 11 cm/s in the septal LV wall.

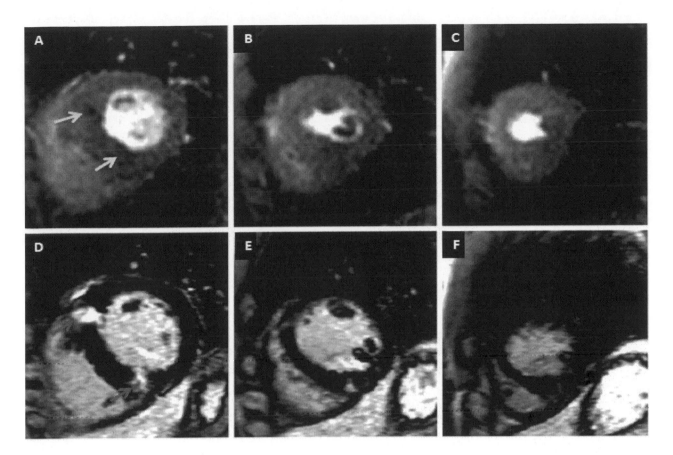

Figure 5: Stress perfusion images (A-C) at basal (A), mid (B), and apical(C) ventricular levels. Focal areas of hypoperfusion are seen (yellow arrows) in the basal anteroseptal and inferoseptal walls. Late gadolinium enhancement (LGE) images (D-F) demonstrate focal myocardial fibrosis (red arrows) predominantly at basal level in the basal anteroseptal and inferoseptal walls.

Subsequent investigations included an exercise stress test with the athlete completing 21 minutes of the Bruce protocol (19.1 METS) corresponding to an oxygen consumption of 67 ml/kg/min. The heart rate (91% predicted maximum), BP response (systolic BP rising from 98 mmHg to 168 mmHg at peak exertion) to exercise was normal and there was no evidence of cardiac dysrhythmias on exercise or on the 24-hour Holter monitor.

The abnormal resting ECG, asymmetric septal hypertrophy and non-dilated LV cavity raised suspicion of HCM. However, the normal indices of diastolic function and supra-normal functional capacity favoured 'athlete's heart'. Consequently, CMR was performed using standardised imaging protocols, demonstrating asymmetrical hypertrophy predominantly affecting the basal and mid anteroseptal and inferoseptal walls (maximum wall thickness, 17mm). The lateral wall at the same level measured 8.5mm. There was an inducible perfusion defect noted in the regions of maximal hypertrophy (Figure 5A), whilst late gadolinium enhancement demonstrated regions of focal intramyocardial fibrosis in the anterior and inferior basal LV-RV insertion points (Figure 5D).

Accordingly, first-degree relatives were invited for cardiovascular screening, which demonstrated an abnormal ECG and echocardiogram in the mother and sister of the index case. Genetic testing for known mutations encoding sarcomeric contractile proteins proved positive for a mutation in the MYBPC3 gene encoding myosin-binding protein C in all three individuals.

Discussion

Previous cases of endurance athletes with HCM have been reported [8], however, this is the first case of a genetically proven diagnosis in an individual able to perform such high levels of ultra-endurance exercise for over 25 continuous years. This case demonstrates the challenges faced when evaluating athletes with an abnormal ECG or LV hypertrophy on echocardiography and highlights the importance of systematic evaluation (including CMR, maximal exercise stress testing and where possible, genetic testing) to determine whether such changes represent physiological adaptation or pathological phenomena.

The ECG presented here has numerous features compatible with cardiac adaptation to exercise, including voltage criteria for LVH, ST-segment depression, a prolonged PR interval and inverted T-waves [5, 9, 10]. However, the repolarisation abnormalities seen in this case (T-wave inversion in V5-6) are nearly always associated with cardiomyopathy and should always be interpreted with suspicion. Whilst echocardiography demonstrated a LVWT is 14mm at the septum, it is well established that a minority of Caucasian athletes (<2%) also demonstrate physiological LVH between 13-16mm [6, 7, 11]. However, physiological LVH is typically associated with LV cavity dilatation of 55-65 mm. Hence, the LV cavity size of 44mm in this case is unexpectedly reduced and typical of the disparity seen in individuals with HCM. Typical additional echocardiographic features suggestive of HCM include systolic anterior motion of the mitral valve apparatus, LV outflow tract obstruction and impaired diastolic function, all of which were absent in this case.

In athletes, diastolic function is normal due to a compliant LV and prolongation of diastole allowing for sufficient LV filling and maintenance of a high stroke volume at rapid and sustained heart rates. In individuals with HCM, myocyte disarray and myocardial fibrosis lead to myocardial stiffening and impaired LV relaxation manifesting as impaired early, passive filling of the LV with reversal of the E:A ratio (<1) and prolonged E deceleration time (>240ms). Assessment of longitudinal function on tissue Doppler may demonstrate lower early diastolic velocities, with E' <9cm/s and E/E' ratio of >12; all of which were not observed in this case. After initial investigations, we were faced with an asymptomatic ultra-endurance trained athlete able to perform remarkable levels of aerobic exercise over many years, but with features on ECG and echocardiogram suggesting a morphologically mild case of HCM. Subsequent investigation with exercise stress testing further demonstrated an astounding exercise capacity favouring 'athlete's heart' over HCM.

Given the diagnostic uncertainty in this athlete, this case study also presents evidence for the mandatory inclusion of CMR in evaluating individuals suspected of harbouring a cardiomyopathy. However, it is intriguing to note that despite the observed asymmetric septal hypertrophy and focal intramyocardial fibrosis in the anterior and inferior basal LV-RV insertion points, we postulate that the athlete's ability to sustain prolonged periods of ultra-endurance activity is due to his normal relaxation and passive LV filling properties, a prolonged diastole and normal early diastolic velocities augmenting stroke volume.

Due to the abnormal CMR scan and the abnormal cardiovascular evaluation of first-degree relatives that raising the suspicion of familial disease, requested genetic testing confirmed a diagnosis of HCM. In our opinion, whilst the genetic test confirmed HCM, it was the abnormal CMR scan taken together with the abnormal ECG and echocardiogram in the index cases sister and mother that finally offered the suspicion of HCM over athlete's heart. Whether the genetic test returned negative, it would not have changed the outcome for this athlete.

Pre-participation screening data from Italy, incorporating the 12-lead ECG, suggests that the incidence of sudden death from HCM may be reduced through earlier identification and subsequent disqualification of affected athletes from competitive sport [12]. Although a risk stratification algorithm for HCM is in existence, extrapolation of such data to an athletic milieu with associated high circulating catecholamines, acid-base shifts and electrolyte imbalances is unrealistic. Based on these considerations the exercise guidelines for this heterogeneous disorder are homogenous and conservative [13], and include athletes who may genuinely be at low risk of fatal cardiac events, as in this particular case. Sporting disqualification from all high-intensity ultra-endurance activity was discussed with the athlete. Education was given to the athlete regarding the risks of ultra-endurance exercise. Accordingly, the athlete is required to undergo a comprehensive yearly cardiovascular examination. However, the athlete continues to compete in ultra-endurance running events; and with 2 years of follow-up data, he remains asymptomatic without any significant cardiac changes.

Conclusion

This case study reports an asymptomatic male athlete with 25 years of ultra-endurance competition, with genetically confirmed HCM phenotypically manifesting with LVH, a small LV cavity together with repolarisation abnormalities suggestive of HCM. Despite documented asymmetric hypertrophy and focal myocardial fibrosis in the basal anteroseptal and inferoseptal walls, it is suspected that the athlete is able to run ultra-marathons due to a compliant LV with normal diastolic and systolic parameters, which is able to augment stroke volume. In conclusion, rare as they might be, a minority of HCM patients are capable of life-long careers in ultra-endurance exercise.

References

1. Corrado D, Basso C, Leoni L, et al. Three-dimensional electroanatomic voltage mapping increases accuracy of diagnosing arrhythmogenic right ventricular cardiomyopathy/dysplasia. *Circulation* 2005;**111**:3042-50.
2. Elliott P, Andersson B, Arbustini E, et al. Classification of the cardiomyopathies: a position statement from the European Society Of Cardiology Working Group on Myocardial and Pericardial Diseases. *Eur Heart J* 2008;**29**:270-6.
3. Maron BJ, Shirani J, Poliac LC, et al. Sudden death in young competitive athletes. Clinical, demographic, and pathological profiles. *Jama* 1996;**276**:199-204.
4. Van Camp SP, Bloor CM, Mueller FO, et al. Nontraumatic sports death in high school and college athletes. *Med Sci Sports Exerc* 1995;**27**:641-7.
5. Basavarajaiah S, Boraita A, Whyte G, et al. Ethnic differences in left ventricular remodeling in highly-trained athletes relevance to differentiating physiologic left ventricular hypertrophy from hypertrophic cardiomyopathy. *J Am Coll Cardiol* 2008;**51**:2256-62.
6. Whyte GP, George K, Sharma S, et al. The upper limit of physiological cardiac hypertrophy in elite male and female athletes: the British experience. *Eur J Appl Physiol* 2004;**92**:592-7.
7. Pelliccia A, Maron BJ, Spataro A, et al. The upper limit of physiologic cardiac hypertrophy in highly trained elite athletes. *N Engl J Med* 1991;**324**:295-301.
8. Maron BJ, Wesley YE, Arce J. Hypertrophic cardiomyopathy compatible with successful completion of the marathon. *Am J Cardiol* 1984;**53**:1470-1.
9. Wilson MG, Chatard JC, Hamilton B, et al. Significance of deep T-wave inversions in an asymptomatic athlete with a family history of sudden death. *Clin J Sport Med* 2011;**21**:138-40.
10. Wilson M, Chatard JC, Carre F, et al. Prevalence of Electrocardiographic Abnormalities in West-Asian and African Male Athletes. *Br J Sports Med.* 2012; **46**:341-347.
11. Rawlins J, Bhan A, Sharma S. Left ventricular hypertrophy in athletes. *Eur J Echocardiogr* 2009.
12. Corrado D, Basso C, Pavei A, et al. Trends in sudden cardiovascular death in young competitive athletes after implementation of a preparticipation screening program. *Jama* 2006;**296**:1593-601.
13. Maron BJ. Contemporary insights and strategies for risk stratification and prevention of sudden death in hypertrophic cardiomyopathy. *Circulation* 2010;**121**:445-56.

Commentary: *Keith George*, Liverpool John Moores University, UK.

This is a very interesting and well-documented case study of a well-trained ultra-endurance athlete who entered a cardiovascular screening process on the basis on initial presentation of a cardiac murmur. The detailed investigation, interpretation, follow-up etc. are highly informative in a complex case and also evidence of a high-quality, inter-disciplinary investigation by an appropriate group of experts. This latter point is vital when looking at any difficult differential diagnosis of cardiovascular disease and/or the athletic heart syndrome.

Inherited or congenital cardiac diseases can increase the risk of sudden cardiac death and this is exacerbated somewhat in young, healthy, exercising groups. Screening of young athletes for cardiac diseases that may place them at risk of a serious cardiac event is currently a "hot topic" of debate within clinical and sporting circles. The potential to identify someone with disease who could be at increased risk is often weighed against financial costs, problems with false-positive and false-negative diagnoses as well as who should be performing these procedures.

Despite on-going debate, there are some "beliefs" that have been generated about the cardiac screening process and the co-existence of a cardiac disease with the athletic heart. One of these has been, as the authors noted, the fact that endurance training and competition, over many years is rarely observed in cases of documented Cardiomyopathy. The thought has been that if people have Cardiomyopathy they are self-selected out of events that require sustained and high levels of cardiac performance/output. Hence, the statement that many sudden cardiac deaths occur in team sport players where good technique and skill can lead to elite status without the need for exceptional cardiac capacity or fitness. The one interesting point here is that whilst the training load and maximal aerobic capacity is high it would be interesting to note level of attainment in the sport. That said this is a great example of a case that doesn't fit the "rules" or "general beliefs" and as such is a wonderful and timely reminder to all clinicians and scientists to follow due-procedures in all cases.

In this specific case, the personal history, excellent aerobic fitness and general cardiovascular risk, point to an athletic heart. It is the "red flags" raised by the ECG that prompt further study. The small LV cavity (increase septal wall-to-chamber ratio) is also important although function is normal. It is interesting that the smaller free wall has a normal E'/A' image but that the hypertrophied septum has a reversed E'/A'. The MRI data is fascinating but the case is largely concluded with verification by gene analysis and family member study. A salutary lesson of excellent procedure and process for all involved in cardiac screening!

CHAPTER 4

Restoration of Knee Anatomy, Biomechanics and Function by Meniscal Allograft Transplantation

*Ian D McDermott, MB BS, MS, FRCS(Orth), FFSEM(UK)*London Sports Orthopaedics, london@sportsortho.co.uk and Centre for Sports Medicine and Human Performance, The School of Sport and Education, Brunel University, UK

Vignette

A 36-year-old otherwise fit and healthy man presented with a 10-year history of problems with his left knee. His initial injury (in his mid-twenties) had been a tear of the medial meniscal cartilage from a twisting injury to his knee, sustained during a game of (non-professional) football. His knee injury had been treated (elsewhere) 3 months post-injury with a knee arthroscopy (keyhole surgery), with trimming of the torn meniscus (partial meniscectomy). The patient then underwent a program of physiotherapy rehabilitation, and he did well, returning to full function and with no significant symptoms. He reported that he was able to walk normally, train in the gym and he had been able to return to playing football.

Ten years after the original injury, the patient then presented with an 18-month history of gradually increasing knee symptoms and decreasing function. He complained of pain around the medial side of his knee during and after any kind of exercise or after long walks. The patient reported that he no longer felt able to play football with his children or run at all. He reported that doing this significantly aggravated his knee symptoms. Clinical examination revealed mild tenderness around the medial joint line and a small joint effusion. Magnetic resonance imaging (MRI) scanning showed evidence of a previous sub-total medial meniscectomy plus thinning of the articular cartilage in the medial compartment (on the surface of the medial femoral condyle and the medial tibial plateau).

The patient underwent an arthroscopy (keyhole surgery) of his knee. This confirmed the almost complete absence of a medial meniscus, and so the ragged meniscal rim was tidied up. The articular cartilage in the medial compartment was found to be slightly soft, rough and thin, but there were no areas of full-thickness chondral damage at all.

Post-operatively, a number of different options for treatment were discussed in detail with the patient. The patient decided that he wanted to proceed with soft tissue reconstructive surgery in his knee. Therefore, 8 weeks after the arthroscopy the patient underwent replacement of the missing medial meniscus by meniscal allograft transplantation. This was performed using a fresh-frozen, non-irradiated, chemically sterilized meniscal allograft, inserted using an arthroscopic technique with bony fixation of the anterior and posterior meniscal horns via transtibial tunnels plus peripheral meniscal suturing (Figure 1).

Post-operatively, the patient's knee was protected in a brace, locked at 0 to 90⁰ flexion, and non-weight bearing for 6 weeks, followed by a 6-week course of intensive physiotherapy exercises. By 3 months post-op, the patient was using an exercise bike, and by 9-months post-op he was allowed to resume heavy weights, twisting and impact exercise.

At latest follow-up, 2 years post-transplantation, the patient reported occasional mild aching in the knee after exercise, but no actual pain or swelling anymore. The patient no longer felt that he had any significant functional restrictions and he was able to return to full exercise, including running and playing just gentle football with his children.

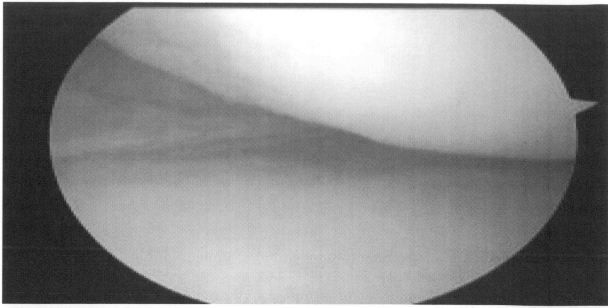

ELEMENTPHYSICAL

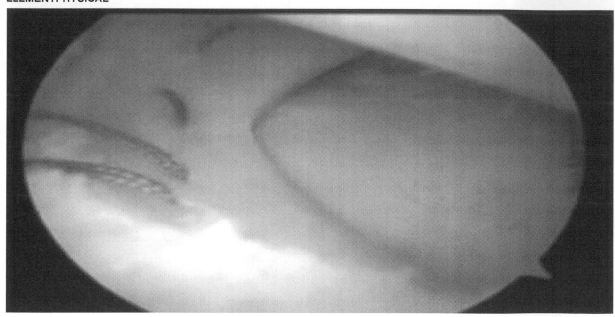

Figure 1. A. Medial compartment of patient's knee with medial meniscus missing.
B. Same view but after implantation of a medial meniscal allograft.

Meniscal Allograft Transplantation

The menisci of the knee (Figure 2) are important cartilage load sharers and shock absorbers in the knee, and they have also been shown to have roles in joint lubrication and nutrition of the articular cartilage as well as being secondary stabilizers of the joint, particularly in the ACL-deficient knee (1). Loss of meniscal tissue (meniscectomy) has been shown to increase the subsequent risk of secondary osteoarthritis in the knee by a factor of 14 when viewed 21 years later (2). Therefore, wherever feasible and appropriate, meniscal repair rather than excision should be performed for symptomatic meniscal tears, although repair is only actually feasible in about 25% of meniscal tears, at best.

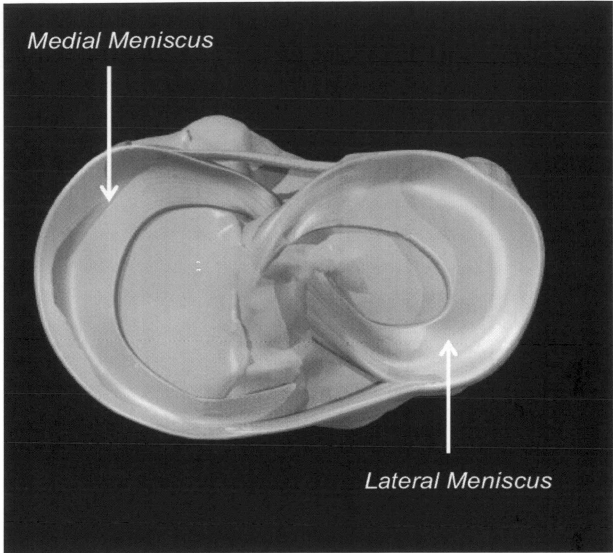

Figure 2. The menisci of the knee. Viewed from above, looking down onto the top of the tibial plateau of a left knee, with the anterior aspect at the bottom.

Human meniscal transplantation using cadaveric allograft tissue was first reported in Germany in the late 80s by Milachowski (3), and since then it had gradually gained acceptance and popularity, particularly in some areas of Europe and in the United States. Meniscal transplantation has been demonstrated to be an appropriate treatment option for those patients suffering knee symptoms and/or early degeneration secondary to previous loss of a meniscus, with studies reporting decreased pain, increased function and success rates in the region of about 85% at 5-year follow-up (4). Meniscal transplantation in patients with more severe degeneration in the knee has been shown to delay the subsequent requirement for knee replacement surgery (5), and meniscal transplantation studies in animals have demonstrated a reduction in the development of arthritis compared to meniscectomized, non-reconstructed controls (6).

The author developed a particular specialist interest in the field of meniscal transplantation and identified a number of questions that were the subject of specific subsequent study, some of which are outlined below.

Meniscal allograft sizing
A review of the literature emphasized the critical importance of implanting a graft of the appropriate size when performing meniscal allograft tranplantation, but revealed a relative paucity of evidence with respect to the various proposed methods for size matching the graft to the recipient patient's knee.

A study was therefore undertaken, examining 88 cadaveric donor tibial plateaus with medial and lateral meniscal allografts attached intact. Meniscal and tibial plateau dimensions were measured (Figure 3). Linear regression analysis was used to calculate expected meniscal dimensions from each specimen's plateau dimensions.

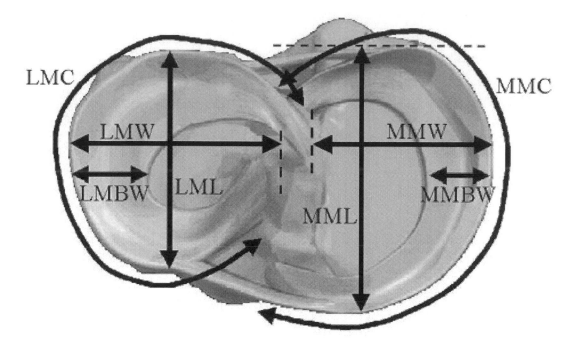

LMC = lateral meniscal circumference   MMC = medial meniscal circumference

LMW = lateral meniscal width   MMW = medial meniscal width

LMBW = lateral meniscal body width   MMBW = medial meniscal body width

LML = lateral meniscal length   MML = medial meniscal length

Figure 3. The various meniscal dimensions that were measured, in trying to size match appropriately a meniscal allograft for a recipient patient's knee, correlating meniscal dimensions to various tibial plateau bony dimensions.

Following the published methodology of Pollard (7), which uses specific medial and lateral tibial plateau width and length measurements, meniscal dimensions could be predicted with a mean error of 5.0% (with a standard deviation (SD) of 6.4%). However, when predicting meniscal dimensions with a simplified method relying only on total bony plateau width, the mean error observed was 6.2% (SD = 8.0%). The difference between the two methods was not statistically significant (8).

The simplified sizing protocol was therefore recommended as an appropriate method of sizing meniscal allografts. Subsequently, this new method is now in the process of being adopted and employed by the Tissue Services Division of NHS Blood Transfusion for the sizing of their meniscal allografts (*Personal Communication*, Ms Clare Moorcroft, NHS BT, 2011).

Meniscal processing and storage methods
Various methods have been employed for sterilizing and storing meniscal allograft tissue. One of the earliest used methods was freeze-drying the grafts; so-called 'lyophylisation'. However, lyophylised grafts were found to shrink significantly after implantation, with adverse clinical results, and this technique was therefore abandoned.

Currently, there are three main methods in popular use for graft processing/storage:
1) fresh-freezing with gamma irradiation;
2) chemical sterilization followed by freezing, and;
3) cryopreservation (controlled-rate freezing in a cryoprotectant solution that maintains viability of some of the donor cells after thawing).

To investigate this further, a study was performed to assess any potential effects that these processes might have on the material properties of human meniscal allograft tissue prior to transplantation. Paired left and right human meniscal allografts attached to tibial plateaus from 22 donors were obtained. Each left and right sample from the pair was randomised into being processed by one of the three different processing techniques. Multiple samples from each meniscal allograft were prepared for tensile and compressive testing. Tensile stress to failure, tensile stiffness, compressive stiffness and creep were determined, with the tester blinded to the group from which the sample came, and the values for each left *vs* right randomised blinded pair of samples were compared. Statistical analysis showed no significant differences in any of the material properties measured between frozen-irradiated, frozen chemically treated, or cryopreserved grafts [presented at the European Orthopaedic Research Society Meeting, Helsinki, Finland, 5th June 2003].

Surgical techniques for meniscal allograft implantation

A variety of different surgical techniques have been described for the implantation of meniscal allografts into the knee, with two main schools of thought: one recommends fixation of the meniscal graft to the tibial plateau with bony fixation plus peripheral meniscal suturing, the other proposes securing the graft with suture fixation only, without bony anchorage.

In order to examine any potential differences in the biomechanical effects of these different surgical techniques, 8 human cadaveric knees were taken and Fuji pressure-sensitive film was used to determine the pressures exerted on the lateral tibial plateau under the following conditions:
1) The intact knee,
2) The knee after total lateral meniscectomy,
3) After lateral meniscal allograft transplantation with bone block fixation,
4) After lateral meniscal transplantation with suture-only fixation.

The results (Figure 4) demonstrated that meniscectomy led to significantly increased peak contact pressures in the lateral compartment. Insertion of a meniscal allograft, using either bony fixation or suture-only fixation, led to a significant reduction in peak contact pressures compared to the meniscectomized state. Fixation of the graft with bony anchorage gave slightly better (lower) peak contact pressures compared to suture-only fixation (9).

Conclusions

This chapter demonstrates clearly how patients can have specific problems from particular knee joint damage from trauma, and how appropriate reconstruction/replacement of the relevant damaged structures can resolve symptoms, improve the biomechanics of the joint and restore function. The surgical technique of meniscal allograft transplantation has been studied extensively, and it has been shown to improve biomechanical loading patterns in the knee joint, decrease patients' pain levels and increase their functional levels, with high reported success/satisfaction rates. Meniscal transplantation is gaining popularity in the UK, but as it is complicated and difficult, it is tending to be undertaken at present only by a small number of specialized knee surgeons with a particular specialist interest in the subject. It is, however, anticipated that meniscal transplantation will gain in popularity as further studies are published and as awareness of the availability of the procedure increases. This chapter emphasizes how many different specific clinical and surgical questions can arise in the treatment of problems such as meniscal deficiency, and how with appropriate study and investigation appropriate answers can be obtained to help guide the clinician in providing the best possible most up-to-date treatments for their patients, based on sound clinical and scientific evidence.

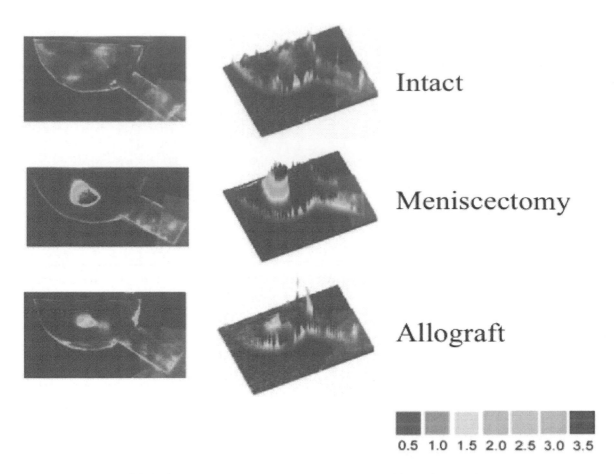

Figure 4. Lateral tibial plateau contact pressures measured with Fuji Prescale pressure sensitive film (left = 2D, right = same data but represented in 3D), comparing TOP – the intact knee, MIDDLE – the knee after removal of the lateral meniscus, and BOTTOM – after lateral meniscal allograft transplantation. (Scale = MPa)

References

1. Masouros SD, McDermott ID, Amis AA, Bull AMJ Biomechanics of the meniscus – meniscal ligament construct of the knee. *Knee Surg Sports Tramatol Arthrosc* 2008;**16**:1121-1132.
2. McDermott ID, Amis AA "The consequences of meniscectomy." *Journal of Bone and Joint Surgery* 2006;**88**-B(12):1549 – 1556.
3. Milachowski KA, Weismeier K and Wirth CJ. Homologous meniscus transplantation. Experimental and clinical results. *Int.Orthop.* 1989;**13**(1):1-11.
4. ElAttar M, Dhollander A, Verdonk R, Almqvist KF and Verdonk P Twenty-six years of meniscal allograft transplantation: is it still experimental? A meta-analysis of 44 trials. Knee Surg Sports Traumatol Arthrosc 2011;**19**:147-157.
5. Stone KR, Adelson WS, Pelsis JR, Walgenbach AW, Turek TJ Long-term survival of concurrent meniscus allograft transplantation and repair of the articular cartilage. *Journal of Bone and Joint Surgery* 2010;**92**-B(7):941-948.
6. Aagaard H, Jorgensen U, and Bojsen-Moller F. Reduced degenerative articular cartilage changes after meniscal allograft transplantation in sheep. *Knee Surg Sports Traumatol Arthrosc* 1999;**7**(3):184-191.
7. Pollard ME, Kang Q, Berg EE Radiographic sizing for meniscal transplantation. *Arthroscopy* 1995;**6**:684-687.
8. McDermott ID, Sharifi F, Bull AM, Gupte CM, Thomas RW, Amis AA An anatomical study of meniscal allograft sizing. *Knee Surg Sports Traumatol Arthrosc* 2004; **12**: 130-135.
9. McDermott ID, Lie DTT, Edwards A, Bull AMJ, Amis AA. The effects of lateral meniscal allograft transplantation techniques on tibio-femoral contact pressures. *Knee Surg Sports Traumatol Arthrosc* 2008; **16**: 553-560.

Commentary: *Henry D.E. Atkinson* FRCS London Sports Orthopaedics,

Joint replacements have been shown to be extremely effective in the management of painful knee arthritis in older patients. Surgery involves removing the knee joint surfaces and typically implanting a metal prosthesis with a polyethylene bearing. This type of surgery is crude and knee replacements are far from being a perfect solution. Functional outcomes are often poor, and implants fail with time, often necessitating further major surgery.

Recently there has been a growing interest in the use of tissue and biological solutions, particularly in younger patients with failing joints (1). The concept of a biological knee replacement involves the regeneration or reconstruction of the original natural joint, with the aim of preventing or delaying the progression of arthritis (1). This approach has huge potential benefits for the active population, with the possibility of restoring knees that have been damaged through injury, and returning them to normal and full function.

One such recent advance has been the use of meniscal allograft transplantation in those patients in whom the native meniscus has been damaged beyond repair. Without a normal functioning meniscus the knee articular surface eventually fails, leading to arthritis. The goal of meniscal transplantation is to cushion and stabilise the injured knee and delay the onset of arthritis. The technique has been extensively studied and has been shown to decrease pain symptoms, improve knee function and improve the biological loading patterns (1-3). Medium-term outcomes for this surgery have thus far been very encouraging with 89% good to excellent results at 10 years and excellent patient satisfaction levels (3,4). Meniscus transplantation has also been successfully combined with cruciate ligament reconstruction, cartilage grafting, and bony realignment surgery (1,5).

Although this ground-breaking surgery is already widely recognised in the US, it is only available in a few European centres, and is slowly gaining in popularity in the UK. The surgery is complex and technically challenging, and a very small number of highly specialised UK knee surgeons currently undertake this surgery.

This case study is an excellent demonstration of the potential benefits of this technique.

References
1. Stone KR, Adelson WS, Pelsis JR, Walgenbach AW, Turek TJ. Long-term survival of concurrent meniscus allograft transplantation and repair of the articular cartilage. *Journal of Bone and Joint Surgery* 2010;**92**-B(7):941-948
2. McDermott ID, Lie DTT, Edwards A, Bull AMJ, Amis AA. The effects of lateral meniscal allograft transplantation techniques on tibio-femoral contact pressures. *Knee Surg Sports Traumatol Arthrosc* 2008;**16**:553-560
3. Cole BJ, Carter TR, Rodeo SA. Allograft meniscal transplantation: background, techniques and results. *Instr Course Lect* 2003;**52**:383–96.
4. Hommen JP, Applegate GR, Del Pizzo W. Meniscus allograft transplantation: ten-year results of cryopreserved allografts. *Arthroscopy*. 2007;**23**:388-393.
5. Packer JD, Rodeo SA. Meniscal allograft transplantation. *Clin Sports Med.* 2009;**28**:259, 83, viii.

CHAPTER 5

Low Energy Availability and Menstrual Dysfunction in an Olympic Speed Skater

Nanna L Meyer[1] and Loretta Cooper[2].
[1]Beth-El College of Nursing and Health Sciences, University of Colorado at Colorado Springs, USA
[2]University of Utah, Salt Lake City, UT, USA

Vignette

A 22-year-old, 2-time Olympic female long track speed skater (all round events: 1500 m; 3000 m; 5000 m) presented with secondary amenorrhea after a 4.5 kg loss of body mass (BM) most likely caused by an increase in training-related energy expenditure secondary to a piriformis injury. Body mass, height, and body composition was 58.2 kg, 167.7 cm and 10% body fat, respectively (BF, measured by 4 skinfold sites; based on Jackson and Pollock (1)). Her lowest BM and BF prior to losing normal menstrual function was 56.8 kg and 8.7%, respectively. She denied pregnancy, contraception, an eating disorder, galactorrhea, heat/cold intolerance and hirutism. She had normal vital functions and normal heart, lung and thyroid functions. Her bone mineral density (BMD; Lunar iDXA, Madison, WI) showed normal Z-scores (lumbar spine: -0.1; proximal femur: 1.0, mean value, left and right side with no side-to-side difference; Figure 1 and 2), although both lumbar and proximal femur values were below what would be expected normal for an Olympic athlete in a weight-bearing sport. Her resting metabolic rate (RMR) was 1,508 kcal/d, measured by indirect calorimetry (Parvomedics, Salt Lake City, UT).

The athlete did not show any eating disorder symptoms and reported no history of oligo- or amenorrhea. Age at menarche was 13 yrs, which occurred after she began skating-specific training at 6 yrs of age. She had a history of exercise-induced asthma and mild symptoms consistent with gastric reflux disease. Her lab values showed prolactin at 6.4 ng/mL, estradiol at 17 pg/mL, TSH at 1.72 uIU/mL, FSH at 3.29 mIU/mL and LH at 1.35 mU/mL.

The athlete's 3-day dietary analysis indicated an energy intake (EI) of 4,145 kcal$\cdot$d^{-1} (71.2 kcal$\cdot$kgBM$^{-1}\cdot$d^{-1}) with adequate macronutrient intakes (carbohydrate: 588 g$\cdot$d^{-1} [10 g$\cdot$kgBM$^{-1}\cdot$d^{-1}]; Protein: 208 g$\cdot$d^{-1} [3.6 g$\cdot$kgBM$^{-1}\cdot$d^{-1}]; fat: 110 g$\cdot$d^{-1} [1.9 g$\cdot$kgBM$^{-1}\cdot$d^{-1}]). Her energy expenditure from exercise (EEE) was 2,453 kcal$\cdot$d^{-1}, and training consisted of cycling, running, and rowing ergometry based on activity records and metabolic equivalents (METs) according to Ainsworth et al. (2). Energy availability (EA), calculated as EI-EEE and expressed per kg of fat-free mass (FFM) was 32.6 kcal$\cdot$kgFFM$^{-1}\cdot$d^{-1}. This was considered a clinically significant change from her initial dietary and EEE analysis when she first started training with the National team as a 19-year-old skater (BM: 63.1 kg; height: 167.7 cm; 14.4 % BF; energy availability: 49.6 kcal$\cdot$kgFFM$^{-1}\cdot$d^{-1}). The higher EA was mainly due to a lower EEE (1,490 kcal$\cdot$d^{-1}) and speed skating-specific exercise modes (i.e., skating, jumping, weight lifting, running, cycling). As a strategy to try to resume menstrual function, the athlete was advised to correct low EA by increasing EI slightly; which coincided with the resumption of speed skating-specific exercise modes. She also received 10 mg of progesterone, which she was advised to take over the course of 10 days but refused to do so. At 1-year follow up, menstrual function (weight: 58.2 kg; 10% BF) with estradiol at 57 pg/mL, FSH at 4.1 mIU/mL and LH at 2.4 mU/mL had resumed. Her dietary analysis revealed a slight increase in EI (4,310 kcal$\cdot$d^{-1}; 74 kcal$\cdot$kgBM$^{-1}\cdot$d^{-1}) and a reduction in EEE (2,130 kcal$\cdot$d^{-1}) which resulted in EA of 41.6 kcal$\cdot$kgFFM-1$\cdot$d-1.

Discussion

Energy availability is recognized as an important factor determining normal menstrual function in athletes. Energy availability is defined as the energy remaining beyond what is required for exercise training, to cover basic physiologic functions, including thermoregulation, growth and reproduction (3, 4). Low EA can occur due to low EI, high EEE, or a combination of the two (3). Athletes with low EA often present with disordered eating and clinical eating disorders. However, not all athletes with low EA and subsequent menstrual dysfunction have disordered eating patterns. In fact, total daily energy expenditure may simply be too high for female athletes to meet through eating solid food, resulting in under-consumption of energy. What causes this inadvertent low EA is currently unclear. It might be that some athletes lack the knowledge and skill to adjust their eating to meet high-energy demands from

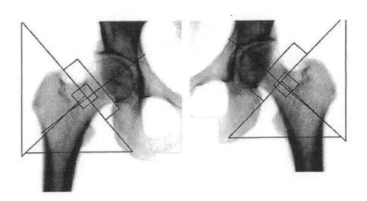

Image not for diagnosis

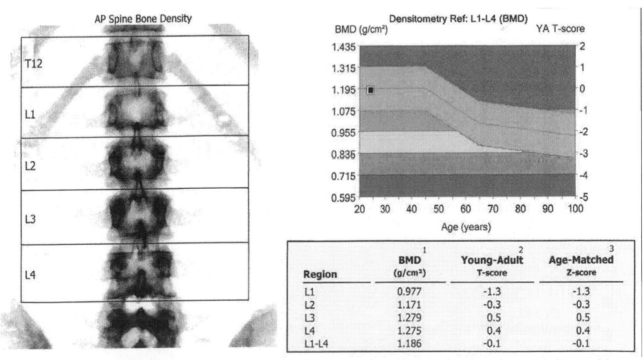

Region	BMD[1] (g/cm²)	Young-Adult[2,7] T-score	Age-Matched[3] Z-score
Neck			
Left	1.176	1.0	1.0
Right	0.994	-0.3	-0.3
Mean	1.085	0.3	0.3
Difference	0.182	1.3	1.3
Total			
Left	1.145	1.1	1.1
Right	1.133	1.0	1.0
Mean	1.139	1.0	1.0
Difference	0.012	0.1	0.1

Figure 1. Bone mineral density at right and left proximal femurs (Lunar iDXA, Madison, WI) in an Olympic speed skater with amenorrhea.

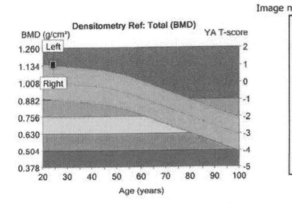

Region	BMD[1] (g/cm²)	Young-Adult[2] T-score	Age-Matched[3] Z-score
L1	0.977	-1.3	-1.3
L2	1.171	-0.3	-0.3
L3	1.279	0.5	0.5
L4	1.275	0.4	0.4
L1-L4	1.186	-0.1	-0.1

Figure 2. Bone mineral density at lumbar spine (Lunar iDXA, Madison, WI) in an Olympic speed skater with amenorrhea.

training. Another possibility is that high-intensity training, travel, and/or competition interferes with the athlete's appetite. Finally, when training several times per day, athletes may lack the time to adequately plan, prepare, and consume food. While it is presently unknown how many athletes struggle with this inadvertent low EA, it is estimated that between 20-30% of elite female athletes are at risk for eating disorders and disordered eating, especially those in thin-built sports (e.g., endurance, aesthetic, weight class; (5, 6). Low EA, regardless of its cause, is a risk factor for menstrual dysfunction.

Menstrual dysfunction varies among sports, with age and training volume. Secondary amenorrhea, as identified in the current case, has been previously reported in 65% of long-distance runners (7), and deteoriates as weekly mileage increases (8). Secondary amenorrhea is defined as the absence of three consecutive menstrual cycles (>90 days) after menarche (3). Amenorrhea is part of the Female Athlete Triad (Triad) and if untreated may result in irreversible bone loss (9). While amenorrhea was historically treated using pharmacological agents, today it is well recognized resolving the underlying cause, most likely being low EA, best treats that hypothalamic athletic amenorrhea. In fact, Loucks and Thuma (10) have shown that low EA at a level of 30 kcal·kgFFM^{-1}·d^{-1}or below is a key factor for disrupting normal menstrual function in previously sedentary, exercising women. Although few data are published on the resumption of menstrual function using re-feeding protocols, animal (11) and human data (12) have shown that the addition of food calories may reverse hypoestrogenism. Cross-sectional data have also shown that athletes with EA below 30 kcal·kgFFM^{-1}·d^{-1} are more likely to be amenorrheic versus athletes whose EA is above 30 kcal·kgFFM^{-1}·d^{-1} (13,14). Amenorrhea is not the only type of menstrual dysfunction. In fact, the recently redesigned Triad graphic shows each Triad component (i.e., EA, menstrual function, BMD) on a spectrum (3). Oligomenorrhea (no cycle for >35 days), anovulatory cycles, and shortened luteal phase are all known as subclinical menstrual disorders occurring in female athletes (3). However, no one occurs preceding the other, and anovulation as well as luteal dysfunction are only diagnosable based on blood and/or urinary measures and are typically not assessed in elite sports.

Athletes with menstrual dysfunction are at risk for irreversible bone loss. Hypoestrognism specifically affects bone turnover by accelerating bone resorption (3). In addition, low EA has been shown to impact bone turnover directly through metabolic, endocrine and substrate (nutrient) effects (15). Compromised bone turnover due to menstrual dysfunction and under-nutrition at young age closes a window of opportunity for female athletes to optimize peak bone mass. While catch-up growth is possible even into the third decade of an athlete's life (16), it is a more common notion that accelerated bone resorption and compromised bone formation impacts bone health in young athletes negatively with little chance of reversibility (9), raising their risk for osteoporosis and fragility fractures later in life. In addition, athletes with menstrual dysfunction have a 2-4 times greater risk for acute stress fractures than their eumenorrheic counterparts, and this risk is magnified in athletes with low BMD (17). Therefore, optimizing bone health through good nutrition, bone-loading exercise and normal menstrual function is of utmost importance in young active women.

Bone mineral density is commonly assessed by dual energy x-ray absorptiometry (DXA) and T and Z-scores provide comparative values to evaluate a person's bone health relative to 1) the mean of young healthy adult women (T-score) or 2) age and ethnicity-matched women/girls (Z-score), with the latter score being a more appropriate value for young, pre-menopausal women. According to the American College of Sports Medicine (3) and the International Olympic Committee Position Statements (18), a T- or Z-score below -1 in an athlete with secondary symptoms, such as hypoestrogenism or nutritional deficiencies, is of concern and considered low BMD, while a T- or Z-score below -2 is defined as osteoporosis. Athletes should have a 5-30% higher BMD than non-athletes (19), especially if involved in weight-bearing sports with high impact. While the ground reaction forces long track speed skaters experience are relatively low (1.5 x BM; (20)), maximal moment values (i.e., sustained loading) especially through the turns are high (21). In addition, speed skating is characterized by vigorous dry-land training, including bone-loading exercise modes (e.g., weight lifting and plyometrics), potentially protecting speed skaters from bone-related injury (22). However, speed skaters with menstrual dysfunction, as shown in the current case, do not fully benefit from this bone-loading stimulus as their BMD gradually becomes compromised. With a few exceptions (e.g., gymnasts (23)), poor BMD is a common threat to athletes with menstrual dysfunction.

Treatment in athletes with menstrual dysfunction must first focus on reversing low EA (3). Manore et al. (24) recommend that female athletes ingest at least 45 $kcal \cdot kgFFM^{-1} \cdot d^{-1}$ during phases of intense training, growth and competition. In addition, meeting nutrient needs necessary for optimal bone health such as adequate calcium, vitamin K, and vitamin D and protein are also targeted. Due to the great risk of vitamin D deficiency in indoor athletes, 25(OH) vitamin D hydroxy should be measured, and supplementation should be based on clinically deficient/insufficient levels. Vitamin D impacts the absorption of calcium and leads to increased levels of parathyroid hormone with subsequent calcium losses from the skeleton (25).

Conclusions

The athlete in this case study presented with secondary amenorrhea due to low EA caused inadvertently by a change in exercise mode, leading to increased EEE secondary to a piriformis injury. The athlete was advised to increase her EI, which coincided with the resumption of speed-skating specific training and a reduction in EEE. As a result, EA increased by nearly 35%. At follow up, the athlete had resumed her menstrual cycle. To reduce the risk of low EA, female athletes should be educated by a sports dietitian on how to adjust EI during phases of higher-energy demands. If lack of appetite, or time to eat, prove to be barriers to optimal fuelling in athletes, liquid meals, flavourful snacks, sport drinks, and bars/gels/blocs may be recommended before, during and after exercise to help maintain optimal EA.

References

1. Jackson AS, Pollock ML, and Ward A. Generalized equations for predicting body density of women. *Med Sci Sports Exerc* 1980;**12**:175-181.
2. Ainsworth BE, Haskell WL, Whitt MC, Irwin ML, Swartz AM, Strath SJ, O'Brien WL, Bassett DR, Jr., Schmitz KH, Emplaincourt PO, Jacobs DR, Jr., and Leon AS. Compendium of physical activities: an update of activity codes and MET intensities. *Med Sci Sports Exerc* 2000;**32**:S498-504.
3. Nattiv A, Loucks AB, Manore MM, Sanborn CF, Sundgot-Borgen J, and Warren MP. American College of Sports Medicine position stand. The female athlete triad. *Med Sci Sports Exerc* 2007;**39**:1867-1882.
4. Wade GN, Schneider JE, and Li HY. Control of fertility by metabolic cues. *Am J Physiol* 1996;**270**:E1-19.
5. Byrne S, and McLean N. Eating disorders in athletes: a review of the literature. *J Sci Med Sport* 2001;**4**:145-159.
6. Torstveit MK, and Sundgot-Borgen J. The female athlete triad: are elite athletes at increased risk? *Med Sci Sports Exerc* 2005;**37**:184-193.
7. Dusek T. Influence of high intensity training on menstrual cycle disorders in athletes. *Croat Med J* 2001;**42**:79-82.
8. Stager JM, Wigglesworth JK, and Hatler LK. Interpreting the relationship between age of menarche and prepubertal training. *Med Sci Sports Exerc* 1990;**22**:54-58.
9. Keen AD, and Drinkwater BL. Irreversible bone loss in former amenorrheic athletes. *Osteoporos Int* 1997;**7**:311-315.
10. Loucks AB, and Thuma JR. Luteinizing hormone pulsatility is disrupted at a threshold of energy availability in regularly menstruating women. *J Clin Endocrinol Metab* 2003;**88**:297-311.
11. Williams NI, Helmreich DL, Parfitt DB, Caston-Balderrama A, and Cameron JL. Evidence for a causal role of low energy availability in the induction of menstrual cycle disturbances during strenuous exercise training. *J Clin Endocrinol Metab* 2001;**86**:5184-5193.
12. Dueck CA, Matt KS, Manore MM, and Skinner JS. Treatment of athletic amenorrhea with a diet and training intervention program. *Int J Sport Nutr* 1996;**6**:24-40.
13. Kopp-Woodroffe SA, Manore MM, Dueck CA, Skinner JS, and Matt KS. Energy and nutrient status of amenorrheic athletes participating in a diet and exercise training intervention program. *Int J Sport Nutr* 1999;**9**:70-88.
14. Thong FS, McLean C, and Graham TE. Plasma leptin in female athletes: relationship with body fat, reproductive, nutritional, and endocrine factors. *J Appl Physiol* 2000;**88**:2037-2044.

15. Ihle R, and Loucks AB. Dose-response relationships between energy availability and bone turnover in young exercising women. *J Bone Miner Res* 2004;**19**:1231-1240.

16. Fredericson M, and Kent K. Normalization of bone density in a previously amenorrheic runner with osteoporosis. *Med Sci Sports Exerc* 2005;**37**:1481-1486.

17. Bennell K, Matheson G, Meeuwisse W, and Brukner P. Risk factors for stress fractures. *Sports Med* 1999;**28**: 91-122.

18. International Olympic Committee Medical Commission. Position Stand on The Female Athlete Triad. 2005.

19. Nichols DL, Sanborn CF, and Essery EV. Bone density and young athletic women. An update. *Sports Med* 2007;**37**:1001-1014.

20. De Koning JJ, de Boer, R.W., de Groot, G., van Ingen Schenau, G.J. Push-off force in speed skating. *Int J Biomech* 1987;**3**:103-109.

21. De Koning JJ, de Groot, G., van Ingen Schenau, GJ. Speed skating the curves. A study of muscle coordination and power production. *Int J Biomech* 1991;**7**:344-358.

22. Meyer NL, Shaw JM, Manore MM, Dolan SH, Subudhi AW, Shultz BB, and Walker JA. Bone mineral density of olympic-level female winter sport athletes. *Med Sci Sports Exerc* 2004;**36**:1594-1601.

23. Robinson TL, Snow-Harter C, Taaffe DR, Gillis D, Shaw J, and Marcus R. Gymnasts exhibit higher bone mass than runners despite similar prevalence of amenorrhea and oligomenorrhea. *J Bone Miner Res* 1995;**10**:26-35.

24. Manore MM, Kam LC, and Loucks AB. The female athlete triad: components, nutrition issues, and health consequences. *J Sports Sci* 25 Suppl 2007;**1**:S61-71.

25. Larson-Meyer E, and Willis KS. Vitamin D and Athletes. *Current Sports Med Reports* 2010.

Commentary *Karen M. Birch*, University of Leeds, UK.

When teaching undergraduates "The Female Athlete Triad" I am often asked whether the athlete always begins their journey along the spectrum of associated disorders with an eating disorder or disordered eating. The Olympic speed skater presented by Meyer and Cooper clearly exemplifies that a change in energy availability can quickly result in disruptions to the hypothalamic-pituitary axis, and thus menstrual dysfunction, without obvious indication of disordered eating. Indeed, the well presented and important message from this case is that a change in energy availability *per se* (expressed per kg of fat free mass) is a catalyst for menstrual cycle disturbance and that this change may be a result of changes in energy intake OR energy expenditure via exercise. Certainly, Williams et al. (1) indicate that menstrual dysfunction with decreased energy availability is more strongly predicted by the magnitude of daily energy deficit rather than losses in body mass. Loucks and Thuma (2) denote a threshold of energy availability (30 kcal·kgFFM^{-1}·d^{-1}) below which menstrual dysfunction is almost certain, whilst Williams et al. (3) very clearly displayed a rapid reversal from secondary amenorrhea in Macaca monkeys running 12 km·d^{-1} once energy availability was increased through re-feeding alone. Furthermore, the rapidity of this recovery was directly related to the amount of energy that was consumed during the period of supplemental feeding.

For an Olympic athlete it is clear that alterations in energy availability can occur via incidental changes in training volume or mode. The skater in this case altered training due to injury from skating specific exercises to a high volume (and thus high kilocaloric expenditure) of aerobic activities such as cycling, rowing and running. The impact of this altered training profile upon energy availability was clearly unintended but provides a weighty exemplar of how even subtle changes in energy metabolism can result in a profound effect upon the reproductive axis and fertility. De Souza et al. (4) provided evidence of a dose-response relationship between measures of energy availability (metabolic and hormonal) and severity of menstrual disturbances such that subtle changes in energy availability *in any physically active women* can still produce dysfunction at the less severe end of the continuum of clinical menstrual disturbances.

The potential impact of both low energy availability and/or menstrual dysfunction upon bone metabolism is the other key message from this case study. Although Z scores were deemed to be normal, they were less than would be expected for an Olympic athlete. From a nutritional perspective it is imperative to recognise that whilst oestrogen deficiency may accelerate bone loss, energy deficiency may decrease the rate of bone formation (5). As such the athlete with deficiency in both is at the greatest risk, but either deficiency should be a priority for prevention, especially as "catch-up" in bone density is ineffectual. For the coach or team awareness of the effect of the combination of inadequate nutritional intake relative to energy expenditure upon the cascade of metabolic consequences for the female performer must be the cornerstone of management.

References

1. Williams NI, Leidy HJ, Legro R, Demers L, Gardner J, Fyre B et al. Predictors of menstrual disturbances in exercising women. In: Endocrine Society. San Diego, CA: Lippincott and Williams; 2005;684.
2. Loucks AB, Thuma JR. Luteinizing hormone pulsatility is disrupted at a threshold of energy availability in regularly menstruating women. *J Clin Endocrinol Metab*. 2003;**88**:297-311.
3. Williams NI, Helmreich DL, Parfitt DB, Caston-Balderrama A and Cameron JL. Evidence for a causal role of low energy availability in the induction of menstrual cycle disturbances during strenuous exercise training. *J Clin Endocrinol Metab*. 2001;**86**:5184-93.
4. De Souza MJ, Lee DK, VanHeest JL, Sheid JL, West SL and Williams NI. Severity of energy-related menstrual disturbances increases in proportion to indices of energy conservation in exercising women. *Fertility and Sterility* 2007;**88**:971-975.
5. Ihle R, Loucks AB. Dose-response relationships between energy availability and bone turnover in young exercising women. *J Bone Miner Res* 2004;**19**:1231-40.

CHAPTER 6

Running Faster: Interventions to Enhance Performance

Jolly Roy
National Sports Institute of Malaysia, Kuala Lumpur, Malaysia

Vignette

The athlete used as the subject of this case study was a 20-year-old male 1500 metre National standard runner. He sought psychological support following disappointing performances in competition. He reported that his performance deteriorated in the final phase of races even though he demonstrated numerous markers of improved performance in training. In the final stages of race he indicated that he felt strong urges to reduce speed coupled with a general feeling of uncertainty over his running ability. An intervention involving multiple methods was used to increase his self-efficacy to manage perceptions of fatigue. Results indicate his self-efficacy increased following intervention and highlight the value of basing an intervention on a detailed examination of the athlete's needs and attempting to address these needs using a solid theoretical base.

Discussion

Due to the physically demanding nature of 1500m running, competitors will need to learn to cope with high levels of perceived exertion (1). One concept proposed to influence the coping process is self-efficacy (2). Self-efficacy is defined as a person's judgment and the belief in one's capabilities to organize and execute the courses of action required for attainment. A number of studies have demonstrated that measures of self-efficacy are predictive of performances in distance running (3, 4). It is argued that self-efficacy beliefs influence the ability to control one's thoughts and emotions, helping to maintain positive beliefs that performance goals will be achieved even when experiencing intense sensations of fatigue. The intervention used with the athlete in the present case study was based on raising levels in the intensity of beliefs in self-efficacy to manage perceived exertion.

The athlete in the present study reported low self-efficacy expectations to manage intense feelings of perceived exertion, particularly those sensations that occur during the last phase of the race. If perceived exertion is considered from a psychological perspective, then the ability to re-interpret sensations of exertion as manageable should enable the athlete to overcome the subconscious impulse to reduce speed. Bandura (2) proposed that self-efficacy beliefs helped individuals cope with discomfort during strength endurance tasks (5), a finding endorsed by Tenenbaum et al (5). Further, as posited in self-efficacy, beliefs mediate thought patterns, affective responses and actions. As individuals base their self-efficacy beliefs on six key elements, then it should be possible to enhance self-efficacy by manipulations one of more of these sources, an approach followed in the present case study.

Six key elements of self-efficacy theory
- i) performance accomplishments,
- ii) vicarious experience,
- iii) verbal persuasion,
- iv) physiological states,
- v) emotional states, and
- vi) imaginal experiences.

In terms of each specific source of self-efficacy, performance accomplishments are proposed to be the most powerful source of efficacy as they are based on personal mastery experiences. In some occasions, the athlete attributed his success to the 'luck of the day' rather than to his own abilities, and therefore, minimized the effects of personal success on subsequent efficacy expectations. Therefore, to try to raise his sense of self-efficacy by ensuring the athletes was successful when he performed. Several strategies were followed.

One strategy was to enter competition where it was likely he would be a successful, typically relatively low-key event. The second strategy was to compete in time-trial races at a shorter distance (e.g., race over

1200 meters) in training. The third strategy was to use mental rehearsal to simulate the demands of the race, something that was particularly useful in time-trial events performed alone. He would rehearse performing in real time with a view to simulating preparation for competition and subsequent performance. This strategy served to provide the athlete with a mental representation of being successful, and further, it also helped the athlete develop a sense of being able to manage intense perceptions of exertion. After time, this practice helped enhance feelings of self-worth, which was evident from his informal spontaneous comments (e.g. "Wow!...I can."). A fourth strategy was to use the coach to reinforce performance successes in his feedback. This feedback was intended to help him refocus his thoughts and feelings. Consequently, the athlete began attributing performance accomplishments to personal factors such as effort and ability, rather than "luck".

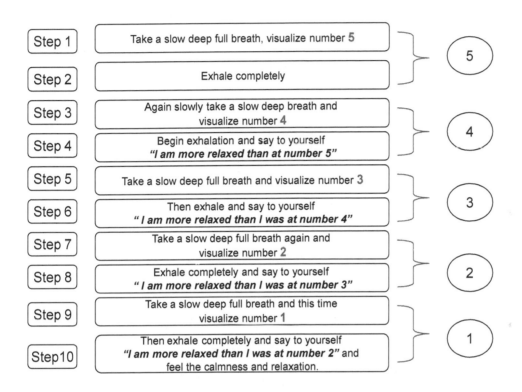

Figure 1. Reverse count breathing (Five to one count)

A fifth strategy intended to increase self-efficacy was through the use of self-talk. Self-talk was directed toward mobilizing effort and increasing his self-efficacy beliefs to persist even though he was experiencing intense physiological discomfort. Self-talk is proposed to represent a form of persuasion in Bandura's (2) model of self-efficacy. In this case study, self-talk was used to help this athlete to overcome his urge to slow down in the latter part of the race (6). The athlete and I explored the verbal cues related to desired actions at specific distance in the final phase of the race and arrived at a series of cues such as 'go' (as he approached the final 300 meters until 200 meters), 'move' (as he advanced from 200 meters to 100 meters) 'fast' (from100 meters) and 'kill it' (for the last 60 meters).

Following the feedback from the athlete that some verbal cues were not helping him manage perceived exertion, we decided to switch the sequence of the cue word with 'move' as he approached the final 300 meters followed by "go". This decision was taken after carefully analysing the athlete's statement such as "my legs don't move when I am close to the final 300 meters". After regularly rehearsing the race mentally using the verbal cues, the athlete's feedback indicated that verbal cues were acting as powerful motivators.

There is a close link between emotions, images and physical sensations (1) and imaginal experiences are proposed to influence self-efficacy (2). Therefore, imagery training specific with the task demands was practiced to help the athlete cope with perceived physiological and emotional states. Athlete's feedback revealed an increased focus while mentally executing the planned sequences (Table 1).

Table 1 Overview of athlete's feedback after imagery (extracts) *Note: Extracts from the last 7 days from a total of 17 sessions*

Sessions	Athlete's feedback after imagery
Session:11	"I Could feel 60 sec for each lap. I was in lane 2 to avoid blocking. After 100 metres I started picking up the race. I said to myself "stay relaxed" Last 500 metres –I pulled and maintained pace. At 300 meters, I used cue-word "move' I could feel body moving faster. I maintained the pace and increased pace. I started feeling little tired. My legs and arms were feeling slightly heavy.
Session:12	"I followed the pace. I could see the other athlete in front of me. I use the cue words in the order push myself at 300 meters. When I started feeling tired…I said to myself "go… go..." at 200 meters "fast" at 100 meters and "kill it" at 60 meters.
Session:13	I said "move" at 300 meters…when I started feeling my legs dragging I said to myself: "go… go…" at 200 meters , "fast" at 100 meters and "kill it" at 60 meters. I felt tired but could overcome the feeling.
Session:14	"I meditated…and flushed out unwanted negative thoughts. I started believing in myself. I knew I could do it. I could see myself leading the race using the cue word "move" at 300 meters used "go" at 200 meters "fast" at100 metres and "kill it" at 60 meters.
Session:15	I Could see the stadium. I Could feel the weather and see my coach .I Could see the crowd… I felt nervous at the start…my thoughts were- "everyone is expecting a gold medal". I felt my heart pounding at the start. My 3 laps were ok with time 60 sec each. Last 300 meters I used the cue words…and I could see me winning with time: 3.47.
Session:16	I was trying to focus on my race. I feel confident. I am trying to Shift my focus from outcomes to performance.
Session:17	I feel confident. I am ready for the race. I can do it.

In the final phase leading up to the competition, I encouraged him to use meditation as a strategy to guard against unproductive thoughts along with slow instrumental music to augment the calming effects. The benefits associated with music are detailed in the conceptual framework suggested recently (7, 8). Researchers (8, 9) have also reported the role of music for calming or stimulating in adjusting psychomotor arousals. Familiar instrumental music (e.g. Solitudes: Woodland flute) stored in a personal portable device helped the athlete manage his arousal and emotions (Figure 1). He used emotive imagery to try to increase arousal for the final part of the race. He imagined the final 60 meters of the race as his 'enemy' and desired to fight with this distance, which probably explains his selection of the verbal cue - 'kill it'. This associative thought was an active coping strategy and concurs with suggestions that an individual's subjective experience of emotion content and its functional meaning in terms of its relation to performance (9).

As inappropriate arousal can result in loss of attention control, I helped teach the athlete how to identify his pre-start preferred arousal states. I used self-report methods to capture optimal arousal levels largely because they are relatively easy to use. A simple time estimation test required the athlete to mentally estimate 10 seconds (on a rest day) using a stop-watch. The athlete then repeated the time estimation prior to the race in simulated competitions. If the estimated time was closer to the rest day value, it indicated a state of readiness. A greater disparity in time estimation from the rest day value (lesser time taken indicated higher arousal and more time taken indicated apathy) helped the athlete to identify his psychological states and self-regulate himself.

Video feedback also served as a valuable aid during intervention in two ways. Firstly, the videotape of the athlete focused on the race as he approached the final 300 metre mark. Feedback was used to initiate discussions focused on the athlete's body movements, particularly, where the athlete was displaying raised shoulders towards the 300 metre mark. Discussions revealed that the athlete was unaware of this effect on his technique, and video feedback served to improve his awareness on the need to control this in training sessions. Secondly, videos of highly successful athletes (role model) were used in an informative manner, to accustom the athlete with the importance of coping with effort. A recent case study was used to illustrate the utility of using video feedback to enhance awareness in applied settings (10). An additional strategy that was initiated was informal group discussions with peer athletes to identify different ways to deal with the difficulties in the final phase of the race. For example, some peer athletes suggested how they dissociated with the discomfort and others how they associated with an internal focus. Finally, as a naive strategy, the athlete's belief that he would get 'strength from God' on the day of competition was never discouraged, considering the cultural context in which he was brought up.

Conclusion

The athlete reported that he could take control of his mind, albeit gradually. Individually tailored intervention provided a structure to develop self-belief, and he progressively made performance accomplishments. Psychological support complemented physical training and this appears to have helped the athlete run faster, and in the process, surpassing a 20-year National record, with an improvement of his personal best time by two seconds.

References

1. Tenenbaum G, Hutchinson JC. A social-cognitive perspective of perceived and sustained effort, 2007. In Tenenbaum G, Eklund RC (Eds.) *Handbook of Sport Psychology*, pp 560-577. John Wiley & Sons, Inc.
2. Bandura *Self-efficacy: The exercise of control*. 1997; New York: Freeman.
3. Martin JJ, Gill, DL. The relationships of competitive orientations and self-efficacy to goal importance, thoughts and performance in high school distance runners. *Journal of Applied Sport Psychology*,1996;7:50-62.
4. Martin JJ, Gill DL. Competitive Orientations, self-efficacy and goal importance in Filipino marathoners. *International Journal of Sport Psychology*, 1995;**26**:348-358.
5. Tenenbaum G, Hall HK, Calcagnini V, Lange R, Freeman G, Lloyd M. Coping with physical exertion and frustration experiences under competitive and self standard conditions. *Journal of Applied Social Psychology*,2001;**31**(8):1582-1626.
6. Zinsser N, Bunker L, Williams JM. Cognitive techniques for building confidence and enhancing performance. In JM Williams *Applied Sport Psychology: Personal Growth to Peak Performance*, 2010; pp316.
7. Karageorghis CI, Terry PC, Lane AM. Development and validation of an instrument to assess the motivational qualities of music in exercise and sport: The Brunel music rating inventory. *Journal of Sports Sciences* 1999;**17**,713-724.
8 Bishop DT, Karageorghis CI, Loizou G. A grounded theory of young tennis players' use of music to manipulate emotional state. *Journal of Sport and Exercise Psychology*, 2007,**29**:584-607.
9. Hanin YL. *Emotions in Sport* Champaign, 2000;IL: Human Kinetics
10. Lane AM. Consultancy in the ring: Psychological support to a world champion professional boxer. In. B. Hemmings & T. Holder (Eds) *Applied Sport Psychology: A Case Based Approach*, 2009; pp51-64.Wiley-Blackwell.

Commentary: Ian Lahart, University of Wolverhampton, UK

The athlete in the above case study experienced deterioration in performance in the final phase of races and experienced feeling a need to reduce speed. The proposed intervention was aimed at reducing the perception of fatigue in the latter stages of a race. Fatigue is highly task-specific and multidimensional. Therefore, it is worth considering both the underlying physiological and psychological mechanisms that may be influence fatigue during a 1500m running event. The 1500 m track running race is a middle-distance event that typically takes about 3.5 and 4 min for elite male and female runners. The energy demands of the working muscles far exceeds the ability of oxidative processes to meet this demand; therefore, there is a high reliance on the anaerobic energy pathways of glycolysis and phosphocreatine breakdown. Physiologically, the successful 1500m performances depend on the ability to provide the required amount of energy at a rate quick enough to maintain the level of muscular activity needed to sustain high speeds for the duration of the event. Fatigue, defined as an inability of a muscle to maintain a required rate of work or speed (1), is an inevitable consequence of 1500m running. It would be advantageous for runners to reduce and/or delay fatigue, or in other words, be able to perform at high speeds for longer.

The exact mechanisms underlying fatigue are still unknown. At the intensity a 1500m runner performs, there is a large depletion in phosphocreatine stores and increased rates of glycolysis. The depletion in phosphocreatine reduces its potential contribution to energy supply. An increased reliance on glycolysis leads to an inevitable increase in the production of hydrogen ions, causing the internal environment of the muscle cell to become more acidic (i.e. fall in pH). The increasingly acidic environment within the cell results in fatigue due to a proposed interference with the chemical reactions involved in glycolysis, thus lowering the capacity of the cell to produce energy this way, and also potential interference with the ability of the muscle cell to contract (1). From a physiological perspective, 1500m runners wishing to improve performance would need to employ training methods that delay the onset of fatigue through an increase in the capacity and rate at which energy is supplied to the working muscles via the anaerobic energy pathways and oxidative metabolism. Performance of intensities at and above the individual athletes $VO_{2\,max}$ would heavily tax anaerobic and oxidative pathways (2). One such training method employed successfully is high-intensity intermittent training, which consists of multiple bouts of short-duration (15 to 180 s) high-intensity ($> VO_{2\,max}$) (3, 4).

The above training methods may help to delay the onset of fatigue; however, it may not directly resolve the problem of how the athlete deals with the sensation of fatigue once it occurs. Recently, Marcora and colleagues (5) suggested that mental fatigue limits exercise tolerance through higher perceptions of effort rather than cardiorespiratory and musculoenergetic mechanisms. While the findings of Marcora and colleagues are controversial, the observations of this current case study appear to be consistent with the idea that high perceptions of effort limit performance. The current case study provokes an interesting idea; assuming the appropriate training stimulus is provided, through psychological training can we teach athletes to improve how they manage intense perceptions of exertion and does this then translate to improvements in performance?

References
1. Maughan RJ, Gleeson M. *The biochemical basis of sports performance*. 2nd ed. Oxford: Oxford University Press; 2010.
2. Hazell TJ, Macpherson RE, Gravelle BM, Lemon PW. 10 or 30-s sprint interval training bouts enhance both aerobic and anaerobic performance. *Eur J Appl Physiol*. 2010 Sep;**110**(1):153-60.
3. Whyte G, British Association of Sport and Exercise Sciences. The physiology of training. Edinburgh: Churchill Livingstone Elsevier; 2006.
4. Esfarjani F, Laursen PB. Manipulating high-intensity interval training: effects on VO2max, the lactate threshold and 3000 m running performance in moderately trained males. *J Sci Med Sport*. 2007 Feb;**10**(1):27-35.
5. Marcora SM, Staiano W. The limit to exercise tolerance in humans: mind over muscle? *Eur J Appl Physiol*. 2009 Jul;**109**(4):763-70.

CHAPTER 7

Coping with Emotions in Motorbike Enduro Racing

Montse C. Ruiz
Department of Sport Sciences, University of Jyväskylä, Finland.

Introduction

The following case study presents an athlete and task-oriented individualized assessment and intervention strategies used in the regulation of the emotional states experienced by a motorbike enduro racer[1]. Enduro racing is a motorcycle sport that takes place on off-road courses. In contrast to motocross racing, enduro drivers race alone during a number of stages where they need to get around trees or other obstacles in a race against the clock. The riders are given specific times when they should arrive at a certain location along a prescribed route, with early or late arrivals resulting in penalties. In general, competitions last about 7 to 8 hours with timed-sections lasting about 30 min.

Vignette

A 27-year-old motorbike enduro racer first contacted me before the summer break because he had concerns about his performance. Jack had eight years of experience having participated in national and international races. During our first sessions, we discussed Jack's accounts of the current situation, his past performance history and identified the concerns to be addressed (1). Jack had displayed good performance during the previous season, having placed high in the National rank. At the beginning of the season, he had signed a contract with a new team who offered him better conditions. While his performance in the first two races was within his expectations, during the third race, he had fallen off the motorbike losing much time and arriving late to the next location, which cost him many places on the rank. His performance on the two following races had not been satisfactory and Jack had gradually started feeling that "competing was not fun anymore." Jack did not have a coach; however, his father accompanied him to the competitions, supporting him and providing reference times to other riders. He mainly highlighted problems with concentrating and blocking out distracters. Such distracters usually come in the form of negative and task-irrelevant thoughts. He was sometimes too worried about the performance of other riders, especially after making mistakes.

Both initial assessment and intervention were based on individualized emotion profiling (2). Jack recalled his best and worst races and identified the idiosyncratic content and intensity of his optimal and dysfunctional emotions (see Figure 1). We discussed the differences between the emotion profiles and analysed emotional experiences associated to his successful and poor races aiming at the identification of more stable emotional patterns. Jack's optimal emotional states included feeling "willing" and "alert" and moderately "anxious." He pointed out that feeling worried or too tense were never helpful states. Usually, when Jack had been performing well in the first timed-sections he felt confident and his performance would improve over the race, however, when he did not have a good start or lost time to his opponents he would feel anxious or tense having many difficulties focusing on riding.

We explored the meta-experiences or attitudes he had towards his emotions and self-regulation strategies. After discussions, it became apparent that Jack did not have effective strategies to cope with adverse situations such as when things did not go well. For example, one of Jack's biggest worries was racing in the rain. When he raced on wet conditions negative thoughts (i.e. *"I am not going to do it well," "I will fall down"*) were noticeable in his thoughts. These thoughts triggered feelings of anxiety and fear of failure, which were associated with general tension in his body which Jack had difficulties controlling.

---

[1] The author has received permission from the athlete to present these elements of the consultancy work.

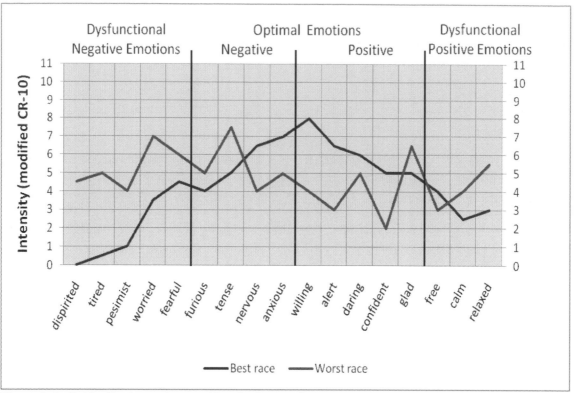

Figure 1. Individualized emotion profiles for best and worst races

Discussion

Emotions are important aspects of human functioning. In sport, emotions are seen as integral parts of an athlete's psychobiosocial state, performance process, and her or his total functioning (3). They are triggered by appraisals of a person's relationships with the environment (4,5), thus, reflecting critical moments of balance or imbalance. Emotion-performance relationships are bi-directional with pre-competitive emotions affecting an athlete's performance while an on-going evaluation of the performance process has an impact of the personal meaning the athlete attributes to the situation, which results in a possible change of quality (content) and intensity of emotions (3).

Based on these assumptions, the plan of action developed in the case described above started with the examination of the routines on the preparation day, and pre-race routines. While there are several self-regulation strategies that athletes can use to prepare for competitions or enhance athletic performance (6), in this individualized intervention the approach was exploring and enhancing strengths and coping strategies that were already available to Jack (7).

We analysed critical race situations with a focus on the personal meaning that they had for Jack. For instance, the change of team was apparently a good opportunity to achieve better results since the conditions and resources (i.e. technical, economic) exceeded those of the previous team. However, this was perceived by Jack as a situation where he "must demonstrate his worth" (his own words) to the team, which raised his expectations and created anxiety and pressure to perform.

Emotional profiling was used to enhance Jack's awareness of his optimal and dysfunctional emotion patterns. We examined the functional impact that pre-race emotions had on his riding performance, and how the on-going evaluation of his performance also affected how he felt. For instance, Jack described an optimal position on the motorbike as somehow relaxed, rarely touching the seat or being almost standing and with the elbows separated from the body (to increase mobility specially in the turns). However, when he was nervous or too tense, he modified the riding position adopting a "safer position" where he would sit closer to the seat or even sit down which made moving more difficult. Such pattern of negative thoughts and feelings of anxiety also altered the way he entered the turns, and used the brakes too early therefore losing time to his opponents. Although he was aware of the potential effects, he did not seem to have a plan of action.

We discussed different dimensions of attention (8) and explored Jack's optimal focus on the moments prior to the start and during the race. I emphasized the importance of recognizing distractions and the ability of shifting from an internal focus such as when he engaged in negative thinking or paid attention to muscle tension, to an external focus of attention (i.e. looking at the terrain). Controlling the thought process and remaining focused on the task seemed to be a concern. Thus, pre-race preparation consisted of a plan to focus on task-relevant cues (previously identified). Since timed-sections are long and many obstacles can be encountered, we also worked on developing a plan to re-focus. During races, the main distracters consisted of negative thoughts, especially after having made a mistake. It was suggested that immediately after he engaged in a negative thought (i.e. "*I am losing too much time*"), he would perform a physical action, which should not distract his driving (i.e. a strong grab of the handlebar). This action would be associated with a more positive thought that he had previously identified (i.e. "*I can do it and am going to demonstrate it*") after which he focused on a task-relevant cue (i.e. "*look -at the terrain- in front*"). Jack practiced focusing and re-focusing plans during training sessions (i.e. in very demanding routes). The aim was to enhance his confidence and provide him with a plan of action to block distracters and focus on task-relevant cues. Other cue words were developed regarding race strategy. Jack was also familiar with relaxation techniques, (9) which he had only used on the day prior to the race. Breathing and short progressive relaxation techniques were gradually integrated into his racing routine to decrease his anxiety at the start of a race.

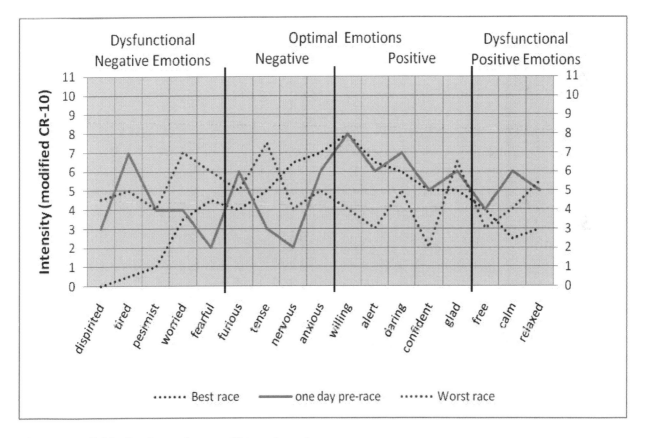

Figure 2. Individualized emotion profiling a day prior to a race

Jack developed emotion profiles, which became a goal for his psychological preparation strategies for the next race. Figure 2 presents the emotion profile he completed one day prior to the race. On the same day, we discussed the profile and contrasted it with his optimal and dysfunctional emotions. He had had an intensive physical training session earlier on the week so he felt very tired. He was also furious because the team had sent the motorbike out a day later than agreed (which never happened before). This meant he had to stay up until late preparing it, to the expense of resting time. We worked on focusing on the aspects he could control, and discussed the possibility of using the additional energy of such feelings to his advantage. Figure 3 presents his emotional profile in the morning of the race, which he completed on the day after. It can be observed that he still felt tired, more nervous than the day before, daring and confident. Although this emotion profile was distant from his optimal emotions especially for positive emotion he reported being nervous but confident and in control. After the race, Jack rated his performance as of high standard (7 on a 0-10 scale).

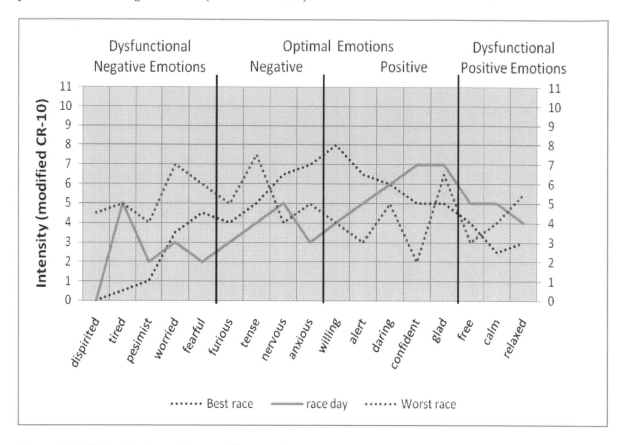

Figure 3. Individualized emotion profiling on the race-day

The effectiveness of the interventions was assessed through post-race discussions and self-reports. He completed emotion profiles after selected races. The aim was to enhance his awareness of optimal and dysfunctional emotional patterns and to identify specific aspects in the preparation of future races. Perhaps, one of the most rewarding aspects of the intervention was knowing that Jack was enjoying the races again, which he attributed to feeling more in control of his emotions and confident because he had plans of action for focusing and re-focusing in difficult racing situations. His performance had been improving at the end of the season (he has now been approached by another successful team). However, this cannot only be attributed to the interventions, since in motor racing there are many other aspects involved (i.e. motorbike, more effective communication with the team). The use of technology facilitated communication in the form of e-mailing and videoconferences through Skype TM, something that allowed frequent communication and discussions of timely issues.

Conclusions

The athlete in the present case experienced problems concentrating and dealing with adverse situations, which he interpreted as leading to underachievement and lack of enjoyment. The intervention plan involved the use of an individualized assessment tool and monitoring of the riders' optimal and dysfunctional feeling states. This tool was used to increase awareness of optimal, dysfunctional and current states and on the functional impact upon performance. Regulation strategies (i.e. relaxation, attentional control, control of thought processes) were used to help the rider regulate his performance-related states. Post-competition evaluation, riders' and consultant's reflections were indicators of the effectiveness of the intervention plan.

References

1. Hanin YL, Stambulova N. Sport Psychology, Overview. In C.Spielberger (Ed.) *Encyclopedia of Applied Psychology*. Vol. 1 (pp. 463-477). Oxford, UK: Elsevier Academic Press; 2004.
2. Hanin YL. Emotions in sport. Champaign, Illinois: Human Kinetics; 2000.
3. Hanin YL. Emotion in Sport: An Individualized Approach. In: C. D. Spielberger (Ed.). *Encyclopedia of Applied Psychology*. Vol. 1 (pp. 739-750). Oxford, UK: Elsevier Academic Press; 2004.
4. Lazarus RS. Progress on a cognitive-motivational-relational theory of emotion. *American Psychologist*. 1991;46:819–834.
5. Lazarus RS. How Emotions Influence Performance in Competitive Sports. *The Sport Psychologist*, 2000;14:229-252.
6. Vealey RS. Mental Skills Training in Sport. In Tenenbaum G, Eklund RC (Eds.) Handbook of Sport Psychology (pp. 287-309). Hoboken, NJ: John Wiley & Sons, Inc; 2007.
7. Robazza C, Pellizzari M, Hanin YL. Emotion Self-Regulation and Athletic Performance: An Application of the IZOF Model. *Psychology of Sport and Exercise*, 2004;5:379-404.
8. Nideffer RM. Attention Control Training. In Singer RN, Murphey M, Tennant LK (Eds.). Handbook of Research on Sport Psychology (pp. 542-556). New York: Macmillan; 1993.
9. Williams J, Harris DV. Relaxation and energizing techniques for regulation of arousal. In Williams JM (Ed.), Applied sport psychology: Personal growth to peak performance (pp. 285-305). Mountain View, CA: Mayfield Publishing; 2006.

Commentary: *Peter C. Terry*, University of Southern Queensland, Australia.

Montse Ruiz presents a detailed case study from the rarely-studied sport of motorbike enduro racing, in which she documents the all-too-familiar trials and tribulations of Jack, a 27 year old professional racer. Montse, a research collaborator of Yuri Hanin's in Jyväskylä, Finland, chronicles Jack's struggles with emotional and attentional control issues in the physically and psychologically demanding world of off-road motorbike racing. Competing alone over long distances in enduro races, Jack has plenty time to think, frequently a curse for athletes. Unsurprisingly, many of his thoughts are random, maladaptive and self-sabotaging. Montse steps the reader through the various stages of support she provided for Jack, helping to transform the anxious, distracted and fundamentally unhappy rider into a more focused, resilient competitor who feels at ease with himself.

Using techniques derived from Hanin's IZOF principles, Montse reports Jack's individualized emotion profiles for best race, worst race and on two occasions leading into an international race. This idiographic approach to understanding emotion-performance links, and subsequent effects upon attentional processes, is a strategy about which neophyte practitioners would be well advised to learn more. Montse's strategy was clearly underpinned by a considerable amount of solution-focused counselling that explored Jack's attitudes towards several aspects of his racing career. Negative attitudes were revised, critical attentional errors were rectified, and the bi-directional relationship between emotions and performance was managed effectively.

The emotion profiles provided a focus for developing Jack's psychological strategies and helped to inform his subsequent discussions with Montse. I was especially taken by the emphasis she placed on Jack regaining his enjoyment of racing. In my practitioner experience, helping athletes to enjoy facing the challenges inherent in international sport is an essential ingredient for sustained success, which is often overlooked in the research and professional literature. Although not discussed directly, oblique reference is made to Jack's broader familial, support, financial, and other life issues. The impact of such matters upon performance is very often profound and it is rare for consultations with athletes to progress far without such issues coming onto the agenda.

Montse also placed considerable emphasis on exploring with Jack the personal meaning he attaches to various aspects of his racing career and how such meaning influences his affective, behavioural and cognitive responses to race situations. This might appear to be standard procedure for those trained as psychologists but it is not a strategy implemented universally by practitioners from a mental training background. Its importance should not be underestimated as a valuable pre-cursor for effective performance-enhancement interventions.

Towards the end of the case study, Montse touches briefly on the increasingly common way of interacting with clients via various forms of Internet communications, which throws up questions of equivalence with traditional face-to-face consultations. Although her discussion on this matter is cursory, she raises a professional issue that is surely destined to become the subject of fervent debate and significant research interest during the coming years. Personally, I have no doubt that her reported style of long-distance interactions with Jack will rapidly become the norm for applied practitioners working with professional athletes.

In summary, Montse Ruiz provides a case study of great authenticity. She describes an interesting series of psychologist-athlete interactions that extend well beyond the stereotypical mental training program. In the process, she offers several insights into how emotion profiling can be used to good effect with clients. Every sport psychologist has a favoured *modus operandi* and, given the extensive research in support of emotion profiling, it is clear that hers is based on a solid, scientific grounding.

CHAPTER 8

Changes on Performance, Mood State and Selected Hormonal Parameters during Under-Recovery: A Case Study of a World-Class Rower

Jaak Jürimäe, Jarek Mäestu, and Toivo Jürimäe
Institute of Sport Pedagogy and Coaching Sciences, Centre of Health and Behavioural Sciences, University of Tartu, Tartu, Estonia.

Vignette

After three weeks of intensified training, a 22-year-old international rower presented with symptoms of extreme fatigue, beyond that witnessed for the rest of the team. The athletes were members of the national quadruple scull team, which has won several medals at World Championships. The symptoms of fatigue in this elite athlete were accompanied by underperformance in rowing-specific all-out performance tests (see Figure 1). During the training camp athlete's´ status was monitored weekly using the Recovery-Stress Questionnaire for Athletes (RESTQ-Sport).

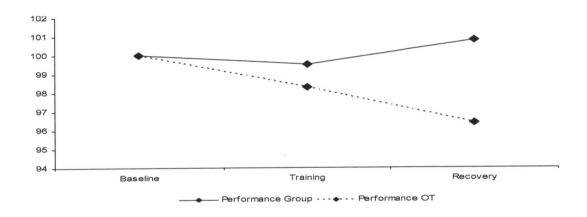

Figure 1. The performance changes (percentage) during the three week high training volume training camp and the following recovery period in overtrained athlete compared to the teammates.

Over the three-week training camp the athlete showed significantly increased General Stress scales of *Fatigue, Fitness/Injury* and also on Sport Specific Stress scale *Burnout/Emotional Exhaustion* scale (Figure 2). These increases in stress experienced by this individual were also seen in his teammates, albeit but to a lesser extent.

Data from the General Recovery scales indicate that the athlete showed significant decreases in *Social Recovery* and *Physical Recovery*. The athlete also reported reduced scores on the Sport Specific Recovery scale; *Burnout/Personal Accomplishment*. However, in athletes that were not diagnosed with Unexplained Underperformance Syndrome (UPS; often previously referred to as Over Training Syndrome, OTS, has been renamed by Budgett et al, 2000) (1) the reduction in Recovery scale scores was rather small. Further, results indicate that all athletes showed decreases in performance after the three-week training camp and were then subjected to a two-week rest and recovery period. Due to the deviation from the scores of teammates on RESTQ-Sport scales, the training of the underperforming athlete was modified slightly during the last week of the training camp such that he had no increase in training volume as previously planned.

After two weeks of recovery all RESTQ-Sport scales were normalized for all athletes except *Fatigue* was slightly increased from Recovery 1 to Recovery 2 in the underperforming athlete. However, during Recovery 1 this athlete showed similar *Fatigue* values as teammates. Despite slightly increased values of

Fatigue the athlete's condition seemed normal compared to teammates. However, when measured again after a two week 'regeneration' period, performance was further decreased in the athlete (a cumulative decrease of 4 % compared with the pre camp value), while other athletes showed an improvement in performance. During the performance tests, the athletes were also subjected to venous blood samples pre, post and 30 min post-test. However, these blood samples were not immediately available for coaches, being analysed later. Following two weeks of recovery the 'over-trained' athlete still had an elevated response in a number of blood analytes such as leptin, insulin, and growth hormone (GH) compared to his own 'normal' values and those of the whole team (Figure 2.). No changes were observed in cortisol or glucose for the three performance tests. Based on these values the athlete continued with another two weeks of reduced training volume. After this, the athlete continued training with the group without any further complications. These data indicate the importance of specific recovery periods for elite athletes and that RESTQ-Sport can be used to detect an athlete who is a non-responder to the training plan. However, in some cases (e.g. after a recovery period) RESTQ-Sport might overestimate the athlete's condition as demonstrated by a further reduction in performance.

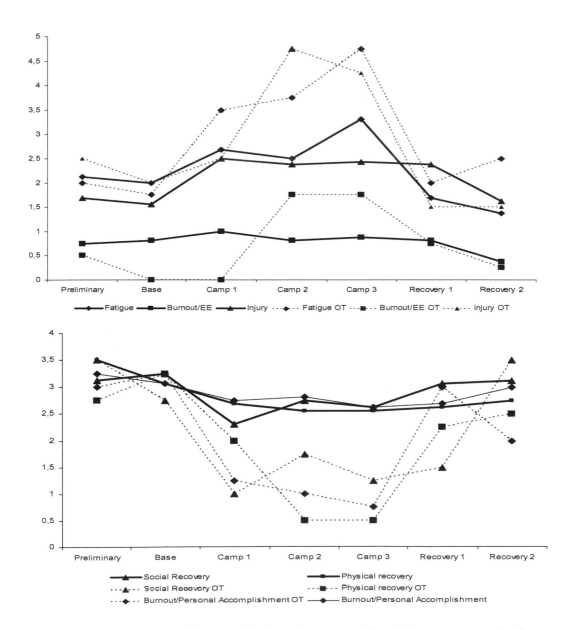

Figure 2. The Stress Scales of the RESTQ-Sport (upper panel) and the recovery scales (lower panel) of the underperforming athlete compared to his teammates during training camp and the following recovery period.

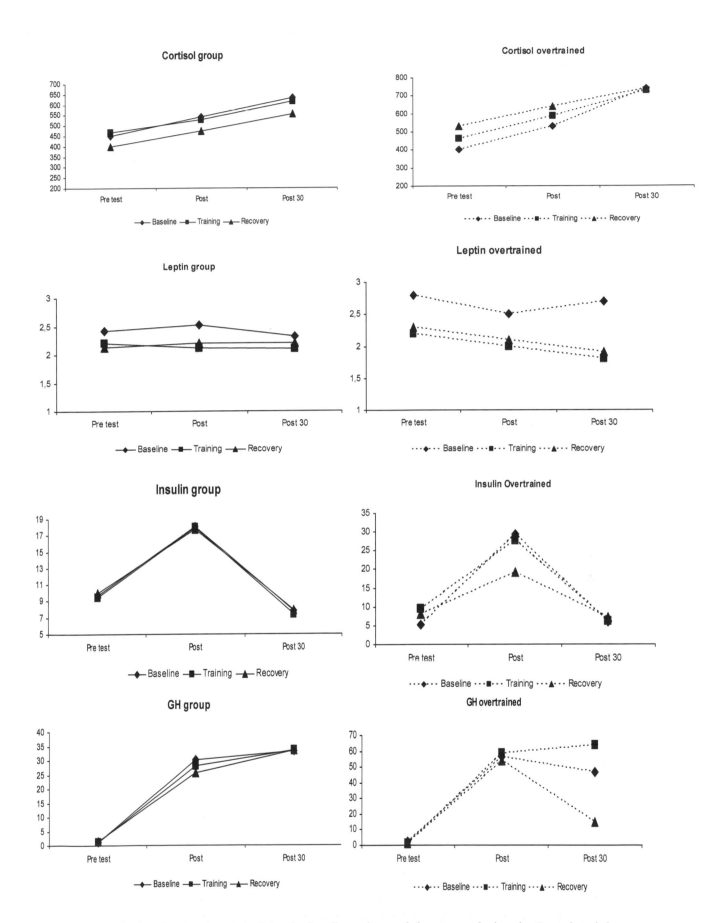

Figure 3. Post exercise changes in cortisol, leptin, insulin and growth hormone during the 3 week training camp and the following recovery period in an underperforming athlete compared to his teammates.

Discussion

It is proposed that self-report data from scores on sport specific performance scale from the RESTQ-Sport should reduce when athletes engage in periods of 'heavy' training. It is worth noting that more objectively measure performance data such as those gleaned from maximal performance tests are not always sensitive enough measures of performance, since their results are subject to athlete motivation (2). However, the extent and content of training often leads to inappropriate training responses and, in the long-term, to 'overtraining syndrome'.

The RESTQ-Sport has often been used during different intervention strategies in athletes from different sport disciplines (3,4,5). An athlete´s recovery-stress status is judged usually by examining scores from an individual profile in comparison to profile scores from team members, assessed under similar conditions. Alternatively, recovery-stress status is compared against an individual database taken from responses to the scale over time (6). Kallus and Kellmann (3) suggested that if more than two RESTQ-Sport scores deviate from the norm, a qualitative interpretation of the patterns provides practical information. Some studies have also found links between RESTO-Sport and other parameters that suggest it has utility in a wider number of areas. For example, a dose response relationship was demonstrated between training kilometres and subscales of RESTQ-Sport (4). Mäestu et al (5) demonstrated a positive relationship between the Stress scales and blood cortisol concentration in blood and between recovery scales and creatine kinase activity.

It is suggested that many of these findings illustrate the importance of recovery in elite athletes preparing for major competitions. Effective recovery from intense training loads has been suggested to be the difference between success and failure in sport (6). More specifically, it is important to diagnose under-recovery as early as possible to prevent further risk of illness or under-performance. Under-recovery was also shown in a study by Mäestu et al. (5) as one of the predictors of failed adaptation to training. In combination with increased *Fatigue* from the general scales and *Fitness/Injury*, *Burnout/Personal Accomplishment* from the sport specific scales (Figure 3) the training for the underperforming athlete was slightly modified (training load was not increased) to prevent failed adaptation. According to Kallus and Kellmann (3) recovery is not just the elimination of stress, but also the process of re-establishing psychological and physical resources for future use. Kallus (7) further postulated that recovery is an individual specific process that occurs over time and depends of the type and duration of stress. In this case-study, the problem for the 'over-trained' athlete probably started with inadequate recovery, since the other team members did not show any marked changes in recovery scales.

UPS may also be characterized by a neuroendocrine imbalance, although the underlying mechanisms are proposed to involve central disturbances at the hypothalamic-pituitary level (8). Athletes suffering from UPS are able to start a normal training sequence, but they are not able to train at a high intensity and in a team environment, this can mean following an individual training plan (9). This might be manifested in a reduced/changed hormonal response to exercise. The literature concerning biochemical parameters in underperforming and 'over-trained' athletes is contradictory mainly due to the duration of training stress and distinguishing between the 'overreached' and the severely underperforming athlete (10). Reduced responses in hypothalamus-pituitary-adrenocortical axis have been shown in athletes with diagnosed UPS / 'overtraining' (9,11,12), with decreased post-exercise cortisol responses in underperforming athletes. Decreased levels of fasting plasma cortisol are considered as a late sign of UPS / 'overtraining'. In addition, a decreased post-exercise response of GH has been reported (13). However, in those with UPS, the metabolic effects of GH are also influenced by a number of other blood bourne analytes (e.g., insulin-like growth factor-1).

There are also leptin receptors in the hypothalamus providing evidence that leptin has effects on hypothalamic neuroendocrine functions (14,15). The metabolic state of the adipocyte is reflected by the leptin concentration since leptin expression is mainly related to flux of energy in the adipocyte (16). High training intensity and volume may create a negative energy balance in the organism due to very high-energy expenditure (14,17) that may result in decrease of leptin response to exercise (18,17) despite relatively low energy expenditure during the exercise itself. However, due to negative energy balance in the organism an additional energy flux in the performance test might be important for an additional decrease in leptin (14,17).

Conclusions

The athlete presented here demonstrated a decreased performance even after a recovery period. However, according to RESTQ-Sport his profile was quite similar to those of his team-mates after the scheduled two week recovery period. Additional blood analysis revealed a decreased post exercise GH, insulin and leptin response, while blood glucose, cortisol and testosterone concentrations remained unaltered. The change in insulin concentration is not often reported in underperforming / 'over-trained' subjects (12). The metabolic effects of insulin include acceleration of anabolic processes in the body by means of growth factors and insulin receptors. Based on the changes in the biochemical data and decreased performance an additional two weeks of recovery was prescribed. Unfortunately, blood biochemical data on this individual were not available after additional recovery, but subsequent performance evaluation of the national team one month after recovery showed that the athlete performed favourably compared with team-mates. In this case, the initial deviation of RESTQ-Sport scales was appropriately considered by the coach, despite the need to modify the training load for that individual for three weeks. However, some care should be taken when interpreting subjective instruments during/after a recovery period, as athletes may be motivated to answer in a way that does not preclude them from future team selection. Hence, a salient lesson here is the need to utilise a range of monitoring tools to prevent under-recovery and underperformance. These include performance measures, subjective ratings and monitoring of some blood analytes.

References

1. Budgett R, Newsholme E, Lehmann M, Sharp C, Jones D, Peto T, Sollins D, Nerurkar R, White P. Redefining the overtraining syndrome as the unexplained underperformance syndrome. *Br J Sports Med.* 2000;**34**(1):67-8.
2. Steinacker JM, Lormes W, Lehmann M, et al. Training of rowers before world championships. *Med Sci Sports Exerc.* 1998;**30**:1158-1163.
3. Kallus KW and Kellmann M. Burnout in Athletes and Coaches. In: Hanin Y, editor. Emotions in Sport. Champaign IL: Human Kinetics, 2000.
4. Kellmann M and Günther KD. Changes in stress and recovery in elite rowers during preparation for the Olympic Games. *Med Sci Sports Exerc* 2000;**32**:676-683.
5. Mäestu J, Jürimäe J, Kreegipuu K and Jürimäe T. Changes in perdeived stress and recovery during heavy training in highly trained male rowers. *Sport Psychol* 2006;**20**:24-39.
6. Kellmann M. and Kallus KW. The Recovery-Stress Questionnaire for Athletes: User Manual. Champaign, IL: Human Kinetics, 2001.
7. Kallus KW. Der Erholungs-Belastungs-Fragebogen [The Recovery-Stress Questionnaire]. Frankfurt: Swets & Zetlinger, 1995.
8. Foster C and Lehmann M. Overtraining syndrome. In: Guten GN editor. Running injuries. Saunders Philadelphia, 1997.
9. Meeusen R, Piacetini MF, Busschaert B, Buyse L, De Schutter G and Stray-Gundersen J. Hormonal responses in athletes: the use of a two bout exercise protocol to detect subtle differences in (over)training status. *Eur J Appl Physiol.* 2004;**91**:140-146.
10. Halson SL and Jeukendrup A. Does an overtraining exist? An analysis ov overreaching and overtraining research. *Sports Med.* 2004;**34**:967-981.
11. Barron JL, Noakes TD, Lewy W, et al. Hypothalamic dysfunction in overtrained athletes. *J Clin Endocrinol Metab.* 1985;**60**:803-806.
12. Uusitalo ALT, Huttunen P, Hanin Y, Uusitalo AJ and Rusko HK. Hormonal responses to endurance training and overtraining in femal athletes. *Clin J Sports Med.* 1998;**8**:178-186.
13. Urhausen A, Gabriel H and Kindermann W. Impaired pituitary hormonal response to exhaustive exercise in overtrained endurance athletes. *Med Sci Sports Exerc.* 1998;**30**:407-414.
14. Jürimäe J, Mäestu J, Jürimäe T, Mangus B and von Duvillard SP. Peripheral signals of energy homeostasis as possible markers of training stress in athletes: a review. *Metabolism.* 2011;**60**:335-350.
15. Steinacker JM, Lormes W, Reissnecker S, et al. New aspects of the hormone and cytokine responses to training. *Eur J Appl Physiol.* 2004;**91**:382-391.

16. Wang Y, Kuropatwinsky KK, White DW, Hawley TS, Tartaglia LA and Baumann H. Leptin receptor action in hepatic cells. *J Biol Chem.* 1997;**272**:16216-16223.

17. Rämson R, Jürimäe J, Jürimäe T and Mäestu J. The influence of increased training volume on cytokines and ghrelin concentration in college level male rowers. *Eur J Appl Physiol.* 2008;**104**:839-846.

18. Jürimäe J, Mäestu J and Jürimäe T. Leptin as a marker of training stress in highly trained male rowers? *Eur J Appl Physiol.* 2003;**90**:533-538.

Commentary *Dr Steve Ingham*, English Institute of Sport, UK

The rowing movement involves cyclical, whole-body, concentric muscle contractions that culminate in force generation at the oar/handle of ~500 to 700 N per stroke (1). In order to repeat this cycle for 5 to 8 minutes the physiological capabilities, and body dimensions, are amongst the largest on the spectrum of athlete sizes, at least in the Olympic programme.

Elite rowers train extensively, approximately 200 to 250 km·week^{-1}. Of this training load, over 85% is performed below the second lactate turnpoint (2). The apparent dichotomy is the disparity between training intensity and racing intensity, that leads to the question asked of rowers, swimmers, cyclists and kayakers, but often not runners; "Why do they train so much and at such a low intensity?"

There are four probable reasons why so much of a rower's training volume occurs at low intensity.
1. The locomotory neuro-muscular recruitment pattern is novel and so, to become autonomous, extensive entrainment is required (3).
2. Injury risk is lower than in other sports, such as athletics, and so, in rowing, limitations to improvement are represented by physiological adaptation, or mal-adaptation, in the form of training/recovery imbalance.
3. There appears to be no further advantage to training at higher intensities (3).
4. Low intensity training is less demanding upon the glycolytic pathways for energy production and thus preserves carbohydrate stores and such training also places less demand upon immune recovery compared with higher intensity work (4). Further, there is comprehensive evidence that elite athletes involved in extensive training 'naturally select' away from 'threshold' type training in order to maximise the volume of low intensity training and optimise the delivery of high intensity training, adopting what could be termed 'polarised training' (5).

During an Olympic cycle with rowing and then with run-based athletics events, I observed the following incidences of musculo-skeletal injury versus cases of over-reaching or unexplained under-performance syndrome (UUPS) (See Table 1). There is an obvious difference between these two sports in the incidences of injury and over-reaching or UUPS and this is, perhaps, explained by the difference in the demands of the events and the specific training required for each. UUPS is prevalent in rowing, but not in athletics; injury is prevalent in athletics but not in rowing.

Table 1. The cases of UUPS or musculo-skeletal injury in rowing vs athletics during a four year Olympic cycle.

	Athlete number (n)	Cases of unexplained under-performance syndrome	Cases of musculo-skeletal Injury
Rowing	40	82	15
	Cases per athlete	2.1	0.4
Athletics (run based events)	25	3	73
	Cases per athlete	0.1	2.9

There are a number of facets to the work of the applied sports physiologist that can be argued that add value.

1. First, understanding why athletes train as they do is, arguably, one of the most important aspects of working with an athlete.
2. Secondly, identifying that the athlete is indeed carrying out the training as the coach intends and that the aims and objectives are met for each session is necessary.
3. Thirdly, assessment of how the athlete is coping and identification of appropriate adaptation is important.

If recovery is measured but the athlete had not adhered sufficiently well to the set programme then, when over-reaching, or UUPS, is identified, this cannot be said to be the fault of the training programme. In this instance, with sight of the data, the athlete can be convinced to adhere to the coach's programme more closely in future. When the two complimentary processes of training and adaptation are monitored closely and compared against progress, then appropriate feedback can be given to improve individualisation and to optimise training.

The search for the 'holy grail', that is a simple marker demonstrating that training-recovery is in balance, is likely to challenge researchers and practitioners for some time. Until that single marker is found, we are left with the challenge of integrating a host of measures to cover differential system breakdown and repair, in order to diagnose a variety of good and bad responses to training. Of the markers currently available, we continually return to simple subjective symptoms of recovery, such as muscle soreness, sleep quality, perceptions of breathing during a standard exercise bout, etc.

The minimalist question, "How are you feeling today?" can help identify potential issues. The athlete's response allows triage;

1. Identification of the specific problem;
2. Assignment to appropriate 'treatment';
3. An improved knowledge of the athlete, and potential to develop a better inter-personal relationship.

However, the real challenge comes when you need to convince the athlete and coach to change the next session as a consequence! A range of evidence, drawn directly from the athlete, that is then collated and shared with the athlete and coach, is probably the most powerful tool in initiating such a change.

References

1. Roth W, Hasart E, Wolf E, Pansold B. Untersuchengen zur dynamik der energiebereitstellung wahrend maximaler mittelzeitausdauerbelastung. *Medizin und sport.* 1983;23:107-114.
2. Steinacker JM, Lormes W, Lehmann M, Altenburg D. Training of rowers before world championships. *Med Sci Sports Exerc.* 1998;**30**:1158-1163.
3. Ingham SA, Carter, H, Whyte GP, Doust JH. Physiological and performance effects of low versus mixed intensity rowing training. *Med Sci Sports Exerc.* 2008;**40**(3):579-584.
4. Halson SL, Jeukendrup AE. Does overtraining exist? An analysis of overreaching and overtraining research. *Sports Med.* 2004;**34**(14):967-81.
5. Seiler K, Kjerland G. Quantifing training intensity distribution in elite endurance athletes: is there evidence for an 'optimal' distribution? *Scand J Med Sci Sports.* 2006;**16**:49-56.

CHAPTER 9

Diagnosis of Asthma / Exercise-induced Bronchoconstriction in Elite Athletes

Pascale Kippelen[1], Claire Bolger[2], & Sandra D Anderson[3]

[1]Centre for Sports Medicine & Human Performance, Brunel University, Uxbridge, Middlesex, UK;
[2] School of Medical Sciences, University of Aberdeen, Aberdeen, UK;
[3] Royal Prince Alfred Hospital and Sydney Medical School, University of Sydney, Sydney NSW, Australia

Vignette

An 18-year-old female international swimmer referred for respiratory symptoms during exercise and under-performance. She trained on average 30 h *per* week (incl. 24 h in indoor chlorinated swimming pools) and has been competing for 11 years. Occasionally, during heavy training sessions, she developed wheeze, chest tightness and a cough, which made her slow down or stop swimming altogether. The symptoms were triggered mostly by exercise and had become worse over the last year. Her coach became concerned about her respiratory discomfort and sought advice on how to treat the symptoms. She had neither a prior medical diagnosis of asthma or of exercise-induced bronchoconstriction (EIB), nor any family history of asthma. She did report symptoms of hay fever.

Results of a baseline spirometry test revealed supra-normal lung function at rest. FVC was 5.07 L (129% pred.), FEV_1 was 4.08 L (117% pred.), FEV_1/FVC was 80%, FEF_{25-75} was 3.85 L.sec^{-1} (99% pred.) and PEF was 7.03 L.sec^{-1} (101% pred.).

A bronchial provocation test was conducted. This involved 8-min of eucapnic voluntary hyperpnea (EVH) of dry air (a surrogate for exercise), with spirometry measurements taken at 2, 5, 10, 15, 20, 30 and 60 min post-challenge. During the test, she reached a ventilation rate of 90 L.min^{-1} (63% of predicted maximum voluntary ventilation, calculated as 35 times baseline FEV_1). After the challenge, she had a significant fall in lung function (Figure 1), with a maximal fall in FEV_1 of 31% at 2 min post-challenge.

The bronchoconstriction was sustained over 20 min and baseline FEV_1 recovered spontaneously within an hour (Table1). A skin prick test revealed significant atopy to house dust mite.

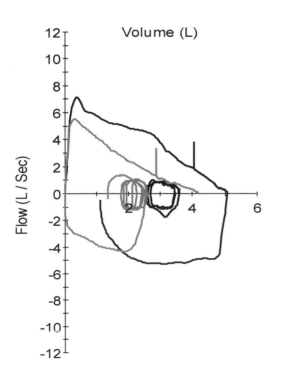

Figure 1. Maximal flow volume loops obtained in an elite female swimmer at baseline (black curve) and 2 minutes after the completion of an eucapnic voluntary hyperpnea test with dry air (pink curve).

Discussion

A high prevalence of asthma and EIB is observed in elite athletes, especially in endurance-trained athletes such as cross-country skiers (14-55%) or long distance runners (15-24%), in swimmers (13-44%), and in athletes training in poor air quality (e.g., 15-19% in ice hockey players) (1). When athletes exercise strenuously in aerobic activities they inhale large quantities of unconditioned air and need to recruit many generations of airways to complete the heating and humidifying process. It has been proposed that EIB is an exaggerated response to airway dehydration in the presence of inflammatory cells and mediators (2). As the rate of water loss from the lower airway is faster than its return, there is an increase in osmolarity of the airway surface lining fluid. According to the osmotic theory of EIB, this increase in osmolarity causes mast cells to release mediators of bronchoconstriction, such as prostaglandins, histamine and leukotrienes (2). In the presence of these mediators, in individuals with a hyperresponsive airway smooth muscle, the muscle contracts and the airways narrow.

Table 1. Forced vital capacity (FVC) and forced expiratory volume in one sec. (FEV$_1$) at baseline and for an hour after eucapnic voluntary hyperpnea (EVH) of dry air in an 18 year old female swimmer with suspected exercise-induced brochoconstriction.

	FVC (L)	Δ FVC (%)	FEV$_1$ (L)	Δ FEV$_1$ (%)
Baseline	5.07	-	4.08	-
Post-EVH				
2 min	4.24	-16	2.82	-31
5 min	4.51	-11	3.05	-25
10 min	4.93	-3	3.52	-14
15 min	4.90	-3	3.61	-12
20 min	4.95	-2	3.62	-11
30 min	5.06	0	3.75	-8
60 min	4.95	-2	4.01	-2

Δ, % change from baseline

Acute dehydration of the airways may also cause epithelial injury and plasma exudation. In elite athletes who exercise frequently at high ventilation, the airway epithelium could be repeatedly injured and the airway smooth muscle chronically exposed to plasma-derived products (including antibodies, cytokines and growth factors). Over time, the process of injury and repair could alter the contractile properties of the airway smooth muscle, rendering it more sensitive (3). Experimental data provide evidence for exercise-induced bronchial epithelial cell injury in athletes (4, 5). Further, the late onset of EIB often observed in elite athletes (6) is in keeping with the development of airway hyperresponsiveness (AHR) following repeated injury to the airway epithelium. In sports performed in poor air quality, noxious agents, such as chlorine derivatives in swimming pools and exhaust fumes from ice resurfacing equipment in ice rinks, could enhance the risk for asthma/EIB (7). Allergic sensitisation has also been identified as a major risk factor for asthma/EIB in elite athletes (8). In the airway injury model it has been proposed that the airway smooth muscle of allergic athletes may become passively 'sensitised' *in vivo* following repeated exposure to plasma derived products (9). This may explain why allergen exposure may increase the likelihood for development of AHR (9).

Despite the fact that specific categories of elite athletes have been identified at risk for asthma/EIB since the late 1990's, undiagnosed and misdiagnosed asthma/EIB is still very common in the sporting world. At the 2004 Summer Olympic Games, asthma/EIB was incorrectly diagnosed in 21% and undiagnosed in 2.6% of the British Olympic squad (10). This may be due to the difficulty in establishing an asthma diagnosis. Whilst GPs often rely heavily on self-reported symptoms, this approach has been shown to be unreliable in athletes (11, 12). The reason being that the characteristic symptoms of asthma, such as breathlessness, cough, phlegm production and wheezing, are commonly reported by athletes during strenuous bouts of exercise; however, they are frequently not accompanied by bronchoconstriction in this population (11). To make the diagnosis even more difficult, some athletes develop AHR, yet remain asymptomatic. Asymptomatic AHR may be linked to the development of tolerance to respiratory symptoms in those individuals performing strenuous physical activity on a regular basis (13).

In order to reduce misdiagnosis of asthma/EIB and misuse of medications in elite athletes, the IOC-Medical Commission introduced in 2011 new recommendations for asthma/EIB treatment in Olympic athletes, based on objective lung function testing (14). The bronchodilator test would provide evidence of reversible airflow limitation at rest in response to inhaling an authorized beta2-agonist or, if lung function was normal, a bronchial provocation test could identify AHR. AHR can be tested using stimuli that act directly on smooth muscle (e.g., methacholine) or stimuli that act indirectly *via* the release of endogenous mediators (e.g., exercise, eucapnic voluntary hyperpnea, mannitol or hypertonic saline). In athletes, the indirect stimuli are preferred to the direct stimuli because the AHR measured in response to indirect stimuli reflects the airway inflammation of asthma and EIB, rather than airway injury (3). Among all the indirect stimuli available, 6 to 8 min of EVH with dry air has been shown to be highly sensitive for identifying those athletes with EIB (12, 15). It is however important to acknowledge that not all athletes respond the same way to the various bronchial provocation tests. In keeping with the possibility of airway injury, cross-country skiers seem particularly hyperresponsive to the pharmacological agent methacholine, but less so to indirect stimuli (16). In contrast, the airways of swimmers are often more likely to respond to the dehydration effect of EVH or of exercise challenges (17). The heterogeneous response of athletes to the different stimuli highlights the importance of selecting an appropriate bronchial provocation test, taking into account: the type of sport, the environment of practice and the presence of an allergic sensitisation.

Once an athlete has been identified with asthma/EIB, s/he will need some form of pharmacological treatment. The pharmacotherapy should be in line with the current international guidelines for asthma management in non-athletes (18). Since some of the drugs may be on the World Anti-Doping Agency list of prohibited substances (19), it is essential that both athletes and doctors check that list before starting a new course of treatment. Currently, all inhaled corticosteroids, but only three inhaled beta2-agonists (salbutamol, formoterol and salmeterol) are allowed for use in and outside of competitions (provided athletes take them in accordance with the manufacturers' recommended therapeutic regimen) (19).
Identification of athletes with asthma/EIB is critical in that drugs usually prescribed for the condition can, if used inappropriately, have some side effects. Such is the case for the widely used inhaled beta2-agonists. Inhaled beta2-agonists are potent bronchodilators and afford good bronchoprotection against EIB. Their regular use however may lead to a worsening of asthma control and development of tolerance (20). Just one week of daily treatment with salbutamol has been shown to result in increased EIB and suboptimal bronchodilator response in patients with EIB (20). Prescription of inhaled beta2-agonists alone is thus strongly discouraged in elite athletes and control of asthma/EIB should be achieved by the use of inhaled corticosteroids for an appropriate period of time.

Conclusion

In the case of our athlete, she was prescribed the inhaled corticosteroid beclomethasone dipropionate (CFC, 200 ug twice daily) and the short-acting beta2-agonist salbutamol (as required). She also started using antihistamine tablets during episodes of hay fever. This treatment provided some relief, with her respiratory symptoms occurring less frequently, enabling her to engage more fully in the training sessions. However, due to the low dosage of inhaled corticosteroid, she continued to report respiratory symptoms during hard workouts and competitions. An increase in the dosage of inhaled corticosteroids was thus recommended (up to 1000 ug of beclomethasone CFC (or equivalent) for a medium daily dose (18)) to treat her airway inflammation and keep her symptoms under control.

An increased prevalence of asthma/EIB has been observed in elite athletes. Early detection of these conditions, using appropriate lung function procedures, is important in order to avoid:
- i) Misdiagnosis,
- ii) Misuse of medications,
- iii) Underperformance, and
- iv) More permanent changes in pathology (airway remodelling).

Given that elite athletes with asthma/EIB have proven to be very successful during the recent Olympic Games (7), proper management of these conditions can be achieved in this population and lead to top performances.

References

1. Carlsen KH, Anderson SD, Bjermer L, Bonini S, Brusasco V, Canonica W, Cummiskey J, Delgado L, Del Giacco SR, Drobnic F. Exercise-induced asthma, respiratory and allergic disorders in elite athletes: epidemiology, mechanisms and diagnosis. Part I of the report from the Joint Task Force of the European Respiratory Society (ERS) and the European Academy of Allergy and Clinical Immunology (EAACI) in cooperation with GA2LEN. *Allergy* 2008; **63**: 387–403.
2. Anderson SD, Holzer K. Exercise-induced asthma: is it the right diagnosis in elite athletes? *J Allergy Clin Immunol* 2000; **106**: 419-428.

3. Anderson SD, Kippelen P. Airway injury as a mechanism for exercise-induced bronchoconstriction in elite athletes. *J Allergy Clin Immunol* 2008; **122: 2**: 225-235.

4. Bolger C, Tufvesson E, Anderson SD, Devereux G, Ayres JG, Bjermer L, Sue-Chu M, Kippelen P. Effect of inspired air conditions on exercise-induced bronchoconstriction and urinary CC16 levels in athletes. *J Appl Physiol* 2011; **111**: 1059–1065.

5. Bougault V, Turmel J, St-Laurent J, Bertrand M, Boulet LP. Asthma, airway inflammation and epithelial damage in swimmers and cold-air athletes. *Eur Respir J* 2009; **33: 4**: 740.

6. Fitch KD. β 2-Agonists at the Olympic Games. *Clin Rev Allergy Immunol* 2006; **31: 2**: 259-268.

7. Fitch KD, Sue-Chu M, Anderson SD, Boulet LP, Hancox RJ, McKenzie DC, Backer V, Rundell KW, Alonso JM and Kippelen P. Asthma and the elite athlete: summary of the International Olympic Committee's consensus conference, Lausanne, Switzerland, January 22-24, 2008. *J Allergy Clin Immunol* 2008; **122: 2**: 254-260.

8. Helenius IJ, Tikkanen HO, Sarna S, Haahtela T. Asthma and increased bronchial responsiveness in elite athletes: atopy and sport event as risk factors. *J Allergy Clin Immunol* 1998; **101: 5**: 646-652.

9. Anderson SD, Kippelen P. Exercise-induced bronchoconstriction: pathogenesis. *Curr Allergy Asthma Rep* 2005; **5: 2**: 116-122.

10. Dickinson JW, Whyte GP, McConnell AK, Harries MG. Impact of changes in the IOC-MC asthma criteria: a British perspective. *Thorax* 2005; **60: 8**: 629-632.

11. Rundell KW, Im J, Mayers LB, Wilber RL, Szmedra L, Schmitz HR. Self-reported symptoms and exercise-induced asthma in the elite athlete. *Med Sci Sports Exerc* 2001; **33: 2**: 208.

12. Holzer K, Anderson SD and Douglass J. Exercise in elite summer athletes: challenges for diagnosis. *J Allergy Clin Immunol* 2002; **110: 3**: 374-380.

13. Turcotte H, Langdeau JB, Bowie DM and Boulet LP. Are questionnaires on respiratory symptoms reliable predictors of airway hyperresponsiveness in athletes and sedentary subjects? *J Asthma* 2003; **40: 1**: 71-80.

14. Anderson SD, Fitch K, Perry CP, Sue-Chu M, Crapo R, McKenzie D, Magnussen H. Responses to bronchial challenge submitted for approval to use inhaled β2-agonists before an event at the 2002 Winter Olympics. *J Allergy Clin Immunol* 2003; **111: 1**: 45-50.

15. Rundell KW, Anderson SD, Spiering BA and Judelson DA. Field Exercise vs Laboratory Eucapnic Voluntary Hyperventilation To Identify Airway Hyperresponsiveness in Elite Cold Weather Athletes. *Chest* 2004; **125: 3**: 909-915.

16. Sue-Chu M, Brannan JD, Anderson SD, Chew N, Bjermer L. Airway hyperresponsiveness to methacholine, adenosine 5-monophosphate, mannitol, eucapnic voluntary hyperpnoea and field exercise challenge in elite cross-country skiers. *Br J Sports Med* 2010; **44: 11**: 827-32.

17. Pedersen L, Winther S, Backer V, Anderson SD, Larsen KR. Airway responses to eucapnic hyperpnea, exercise, and methacholine in elite swimmers. *Med Sci Sports Exerc* 2008; **40: 9**: 1567-1572.

18. Global Initiative for Asthma (GINA). Global Strategy for Asthma Management and Prevention.[Online]. www.ginasthma.org . [Accessed 22/05/2014]

19. The World Anti-Doping Code. The 2010 Prohibited List: International Standard [Online]. World Anti-Doping Agency. http://www.wada-ama.org/en/World-Anti-Doping-Program/Sports-and-Anti-Doping-Organizations/International-Standards/Prohibited-List/ [Accessed 22/05/2014].

20. Hancox RJ, Subbarao P, Kamada D, Watson RM, Hargreave FE and Inman MD. Beta 2-Agonist tolerance and exercise-induced bronchospasm. *Am J Respir Crit Care Med* 2002; **165: 8**: 1068.

Commentary. *John Dickinson*, University of Kent, UK.

The case described here is a common occurrence within the elite athlete community. On completion of the EVH challenge, the swimmer had a maximum fall in FEV_1 of 31% from her baseline measurement. Swimming is a sport that requires high minute ventilation sustained over a significant period during each training session, and takes place indoors, in chlorinated swimming pools. When these factors combine, they create a situation which has the potential to promote airway epithelial damage which, over time, can lead to airway remodelling and increase in bronchoconstriction severity. Given the athlete is only 18 years old she has been diagnosed early in her career as an elite swimmer and will now be able to receive treatment to protect her against future bronchoconstriction.

Detection of athletes with EIB as early as possible is vital to ensure airway health can be maintained but also to ensure performance is not compromised. A study by Haverkamp et al. (1) investigated the effects of inhaled corticosteroids on exercise performance in asthmatic participants. They demonstrated 6 weeks of low dose of inhaled corticosteroids resulted in a 74% improvement in running time to exhaustion, which was accompanied by increases in oxygen consumption, oxygen blood saturation and blood pH. The participants in this study were not elite athletes and they suffered from asthma, as opposed to EIB. However, the study would suggest that treating athletes with inhaled corticosteroids will lead to performance gains. Just how much is difficult to predict as it will depend on the individual, severity of their EIB, and sport they are involved in. Detection of EIB and appropriate treatment gives the athlete every opportunity to maximise their potential in their chosen sport. The reverse can be said in those athletes who remain un-diagnosed and/or sub-optimally treated.

In this particular case study, the swimmer was only tested for EIB once her symptoms became a limiting factor to her training. As mentioned by the authors of the case study, symptoms alone are often inaccurate to pose a diagnosis of EIB in elite athletes (2). It was good practice in this case to perform an indirect airway challenge, such as EVH, to confirm the presence of EIB. It is however, not uncommon for symptomatic athletes (with or without a history of EIB) to have a negative response to indirect airway challenges. In these 'negative' athletes for EIB, a common subsequent diagnosis is inspiratory stridor and/or dysfunctional breathing during high-intensity exercise. There are non-pharmacological treatment strategies that can be used to assist these athletes (3) who would otherwise be prescribed inhaled β_2-agonists to no benefit.

It is well known that many elite athlete compete at international levels of competition without realising they have EIB (4). Therefore, it has been recommended that, where possible, athletes are screened for EIB (5). The screening should involve an appropriate indirect airway challenge test, along with a prior check of previous medical/family history of asthma and, where possible, an assessment of airway inflammation. Follow-up assessments in pharmacologically treated athletes are also worthwhile to examine the attenuation of airway inflammation and bronchoconstriction post-treatment.

References

1. Haverkamp H, Dempsey J, Pegelow D, Miller J, Romer L, Santana M, Eldridge M. Treatment of airway inflammation improves exercise pulmonary gas exchange and performance in asthmatic subjects. *J Allergy Clin Immunol* 2007;**120**:39-47.
2. Rundell K, Im J, Mayers L, Wilber R, Szmedra L, Schmitz H. Self-reported symptoms and exercise-induced asthma in the elite athlete. *Med Sci Sports Exerc* 2001:**33**:208-213.
3. Dickinson J, Whyte G, McConnell A. Inspiratory muscle training: a simple cost-effective treatment for inspiratory stridor. *Br J Sports Med* 2007;**41**:694-5
4. Dickinson J, McConnell A, Whyte G. Diagnosis of exercise-induced bronchoconstriction: eucapnic voluntary hyperpnoea challenges identify previously undiagnosed elite athletes with exercise-induced bronchoconstriction. *Br J Sports Med* 2011;**45**:1126-1131.
5. Holzer, K Brukner, P. Screening of athletes for exercise-induced bronchoconstriction. *Clin J Sport Med* 2004;**14**:134-8.

Part II: The Predictive Model: Providing solutions for events predicted to occur

CHAPTER 10

The Thermoregulatory Challenges faced by the Wheelchair athlete at the Beijing Paralympics in 2008

Victoria L. Goosey-Tolfrey[1] *and Nicholas Diaper*[2]

[1]Loughborough University, School of Sport, Exercise and Health Sciences. The Peter Harrison Centre for Disability Sport. Loughborough. UK.
[2]English Institute of Sport, EIS Performance Centre, Loughborough University, Loughborough, UK.

Acknowledgments: Paul Davies, Jeanette Crosland, Andrew Sommerville, Helen Alfano, Dr Jim House and Dr Stuart Miller (MD).

Vignette

Beijing 2008 was a highly successful Games for ParalympicsGB with the team returning with 42 Gold, 29 Silver and 21 Bronze medals. The team's success was arguably a result of efforts given by not only the athletes who delivered on the field of play, but also by the backroom staff that supported the athletes in their preparation and delivery. The Beijing Games presented a number of unique challenges to our Paralympic athletes of which jet-lag, environment concerns and culture were identified as factors that could be addressed in a four-year programme of support.

With lessons learned from the 2004 Athens Games, a Beijing Acclimatisation Group (BAG) consisting of a multi-disciplinary team of sports scientists and healthcare experts was formed. This team developed individualised support programmes for several of the Paralympic sports with a view to ensuring that each team was best prepared for the challenge of Beijing. This group worked closely with up to 15 Paralympic sports with a variety of different disabilities - amputees, visually impaired athletes, dwarfs, and athletes with cerebral palsy.

The purpose of this case study is to focus on the wheelchair athlete with a spinal cord injury (SCI) to enable the reader to understand how to apply knowledge to special populations with specific issues and needs. The starting point for this programme of sport science support was to review the risk categories of the sporting environment (based upon the nature of the disability, indoor/ outdoor activity and the physical efforts). Interestingly, whilst this 'needs analysis' allowed the BAG team to prioritise areas of support, preliminary data from wheelchair rugby (see Figure 1) resulted in them being moved to the 'high' category (Table 1).

With on-court core temperatures of close to 40°C during wheelchair rugby scrimmages, it was felt that those athletes with a high spinal cord injury who typically display a reduced ability to control body temperature and inability to sweat below the level of injury should review their cooling strategies. The effectiveness of several cooling methods was explored in the build-up to the Games. These methods included hand cooling, cooling hat and neck garments, ice vests and fans (1, 2, 3, 4). Players indicated that using fans combined with a water spray was their preference. The use of fans caused air to flow over the skin and increased convective heat loss, whilst the water spray on the skin acted like sweat and provided similar benefits to sweat evaporation. These methods were demonstrated and trialled by the players prior to Beijing at both the Far East simulation camp in 2007 and subsequent competitions thereafter. Benefits of this work were disseminated across other court based sports (wheelchair basketball and tennis) and track and field athletics.

In terms of fluid replacement, because of the large variations in sweating, drinking recommendations were individualised. This is an area that received considerable attention at both the Far East simulation 2007 and holding camp's in 2008. Trends became apparent based upon disability and sport yet all feedback was individualised rather than taking a generic approach. Results to date have shown an effective hydration strategy was achieved. The analysis of the jet lag data is on-going yet proved a vital tool to assist coaches with the adjustments of training times and intensity of training.

Table 1: Needs analysis of the 'heat stress' risk associated to Paralympic Sporting competition

Heat Stress Level	Event	Indoor/Outdoor
High Risk (Level 3)	Athletics, Cycling,	Outdoor
	Equestrian, Tennis	Outdoor
	Rugby	Indoor
Intermediate Risk (Level 2)	Sailing	Outdoor
	Basketball, Fencing	Indoor
	Swimming, Table tennis	Indoor
Low Risk (Level 1)	Archery	Outdoor
	Boccia, Powerlifting	Indoor
	Shooting	Indoor

Modified from Webborn (1996)(3).

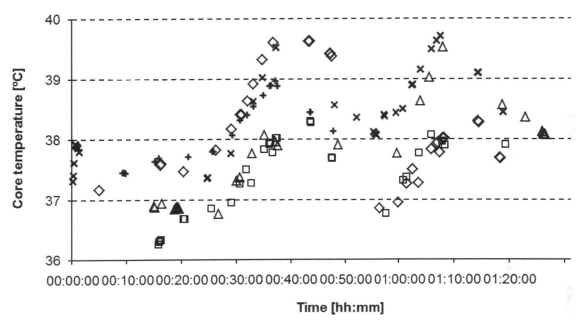

Figure 1. Body temp changes over time for gastric temperature during a wheelchair rugby scrimmage session in a neutral indoor environment (n=5; each shown by a separate symbol).

1. The rationale and interventions

The disruption of autonomic nervous system function resulting from injury to the spinal cord means that individuals with Spinal Cord Injury (SCI) have impaired thermoregulatory control (5, 6, 7). This problem arises as a result of interrupted afferent and efferent input to and from the hypothalamus concerning skin temperature, sweating and vasodilation below the level of injury (8). This leads to a reduced sweating capacity (9) and limited control of blood flow distal to the lesion (10) which could compromise sporting performance and put individuals with SCI at high risk of hyperthermia.

In addition, high temperatures (30-35°C) were expected for the 2008 Paralympics and this was identified as a factor that could significantly impair performance. Therefore, support was provided to determine the physiological and performance effects of a practical cooling strategy that could be implemented during competition within the time-outs. Although earlier work conducted by the ParalympicsGB team (3) had reported the positive effects of wearing an ice vest either before or during a sporting activity in SCI athletes, the application of these methods were not suitable for all Paralympic athletes. It was noted that the ice vest dripped onto the seat of the chair and the vests were difficult to use in a sports setting. Therefore it was necessary to explore alternative cooling methods as seen in the following images.

Figure 2: Cooling strategies adopted by the GB wheelchair rugby team in Beijing 2008.
Images courtesy of Dr Jim House

Hydration monitoring was also conducted at the simulation and holding camps in order to ensure that athletes were sufficiently hydrated prior to training and competition. Maintenance of euhydration has been demonstrated to be beneficial to not only health but also performance (11, 12). Briefly, this involved collection of morning urine samples for the assessment of urine specific gravity using a hand-held refractometer, which has been shown to be an acceptable method of assessing hydration status in athletes (13).

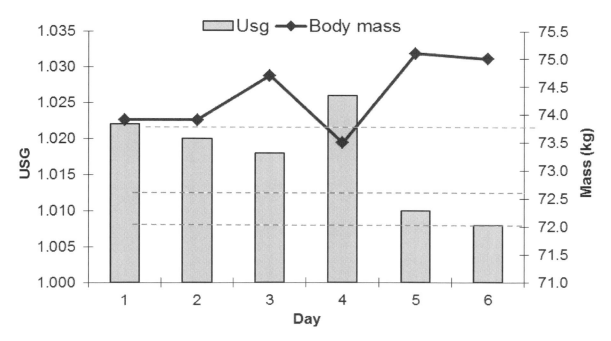

Figure 3: An example of morning hydration status during Macau Holding Camp prior to Beijing 2008 for one wheelchair athlete (note: Urine Specific Gravity (USG) >1.020 = dehydrated; 1.020-1.010 = normal range; 1.010-1.005 = well hydrated and <1.005 = over-hydrated).

Intervention

The results of the hydration monitoring intervention are depicted graphically in Figure 3. This particular athlete was encouraged to aim for urine specific gravity scores of 1.010 each morning prior to training and competition, a score that would indicate being sufficiently hydrated (13). On occasions where this was not achieved the athlete was encouraged to take on fluids prior to training/competition. The athlete was also encouraged to keep a record of fluid consumed on a daily basis as well as to monitor urine colour (13).

2. Evaluation of interventions

The cooling strategy that was developed appeared to have a significant impact on the athletes with respect to their ratings of thermal sensation during competition and the hydration levels of this individual athlete (Figure 3) suggested that fluid replacement strategies were sufficient to ensure adequate hydration levels prior to training/competition once they had overcome the effects of jet-lag.

Although the sport of wheelchair rugby did not medal at the Beijing 2008 Games both athletes and coaching staff stated that they were extremely satisfied with the impact of the support provision. Reflecting on the two to three years spent working with this sport, we think that the introduction of sports science support had a profound and, generally, positive impact on the sport's education, behaviour and actions as highlighted above. A considerable amount of support staff time was spent with athletes and coaches in their own environment at training camps and competitions and this, in retrospect, is viewed as one of the strengths of this programme. We believe that my support in the areas of hydration monitoring and coping with the heat were crucial to ensuring that players were able to train and compete at their optimum levels. This work allowed us to develop an in-depth understanding of the sport and its culture as well as forge strong working relationships with athletes and coaches. We now understand the importance of these factors in delivering effective sports science support and make every effort to incorporate this philosophy into our current practice.

3. Summary

The purpose of the ParalympicGB's 2007 Holding and Simulation 2008 Camps were to provide an optimal preparation environment, free from distraction where athletes can acclimatise and focus, leaving the camp best prepared and ready to deliver personal best performances in Beijing. This goal appeared to be achieved as athlete wellbeing improved throughout the duration of both camps, which suggested that the camp environment and support services provided a positive experience for the athletes, aiding their transit into the Games. It was pleasing to see that athletes adhered to the cooling strategies during preparation and competition, which was underpinned by evidence-based practice. Moreover, athletes were well prepared to maintain hydration despite travel and climate changes, suggesting effective strategies and education processes were put in place.

References

1. Diaper N, Goosey-Tolfrey VL. A physiological case study of a paralympic wheelchair tennis player: reflective practise. *Journal of Sport Science and Medicine*. 2009;**8**:300-307.
2. Goosey-Tolfrey VL, Swainson MG, Boyd C, Atkinson G, Tolfrey K. The effectiveness of hand cooling at reducing exercise-induced hyperthermia and improving distance-race performance in wheelchair and able-bodied athletes. *Journal of Applied Physiology*. 2008;**105**(1).37-43.
3. Webborn ADJ. Heat-related problems for the Paralympic Games, Atlanta 1996. *British Journal of Therapy and Rehabilitation*. 1996;**3**(8):429-436.
4. Webborn N, Price MJ, Castle PC, Goosey-Tolfrey VL. Cooling strategies improve intermittent sprint performance in the heat of athletes with tetraplegia. *British Journal of Sports Medicine*. 2010;**44**(6):455-460.
5. Hagobian TA, Jacobs KA, Kirtali BJ, Friedlander AL. Foot cooling reduces exercise-induced hyperthermia in men with spinal cord injury. *Medicine and Science in Sports and Exercise*. 2004;**36**:411-417.
6. Price MJ, Campbell IG. Effects of spinal cord lesion level upon thermoregulation during exercise in the heat. *Medicine and Science in Sports and Exercise*, 2003;**35**:1100-1107.
7. Chu A, Burnham RS. Reliability and validity of tympanic temperature measurement in persons with high spinal chord injuries. *Paraplegia* 1995;**33**(8):476-479.
8. Hopman MT, Verheijen PH, Binkhorst RA. Volume changes in the legs of paraplegics during arm exercise. *J Appl Physiol* 1993;**75**(5):2079-2083.
9. Sawka MN, Latzka WA, Pandolf KB. Temperature regulation during upper body exercise: able-bodied and spinal cord injured. *Medicine and Science in Sports and Exercise*, 1999;**21**;S132-S140.

10. Theisen D, Vanlandewijk Y, Sturbois X, Francaux M. Cutaneous vasomotor adjustments during arm-cranking in individuals with paraplegia. *European Journal of Applied Physiology*, 2000;**83**:539-534.

11. Montain SJ, Sawka MN, Latzka WA, Valeri CR. Thermal and cardiovascular strain from hypohydration: influence of exercise intensity. *International Journal of Sports Medicine*, 1997;**19**:87-91.

12. Hamilton MT, Gonzalez-Alonso J, Montain SJ, Coyle EF. Fluid replacement and glucose infusion during exercise prevent cardiovascular drift. *Journal of Applied Physiology*, 1991;**71**(3):871-877.

13. Armstrong LE, Maresh CM, Castellani JW, Bergeron MF, Kenefick RW, LaGasse KE, Riebe D. Urinary indices of hydration status. *International Journal of Sports Nutrition*, 1994;**4**:265-279.

Commentary: *Mike Price*, Coventry University, UK.

The account by Goosey-Tolfrey and Diaper provides a unique insight into the support work provided to Paralympic teams prior to their most successful Games to date. Beijing presented concurrent challenges of environmental stress, crossing of times zones and cultural differences. For many able-bodied athletes these would be challenge enough, however, where Paralympic athletes are concerned there is the additional challenge of how disability specific factors may be accentuated in extreme environments. The authors focused specifically upon the area of athletes with spinal cord injury. As these athletes are often considered to demonstrate thermal dysfunction and a greater risk of heat injury (1) this group has received, relative to other disabilities, considerable applied interest in the literature (2, 3), but many unanswered questions remain.

The authors note that prior to Beijing there was a four year multidisciplinary plan resulting in the formation of the Beijing Acclimatisation Group (BAG) and implementation of environmental simulation camps, thus highlighting that Paralympic - and Olympic - support is important, not just for the year of the Games, but begins at the start of each Olympic cycle. An important aspect of the support was that the athletes with tetraplegia, who demonstrate the greatest thermal dysfunction, competing in wheelchair rugby, were placed within the greatest risk category even though their wheelchair rugby events were held indoors. This highlights the heat loss difficulties in these athletes and the potential disruption to performance. Furthermore, athletes who may not be competing outdoors still need to be mindful of enduring environmental stressors when transferring to or between venues or training sessions.

The authors demonstrated how laboratory based information regarding cooling techniques was applied to competitive scenarios in order to determine effective cooling strategies. Only from doing this were they able to determine the practical advantages and disadvantages of each technique. Such a process clearly illustrates that what works in a laboratory may not necessarily be effective in the field. By following this approach, the need for individualised hydration strategies was realized requiring athlete education and engagement. Furthermore, developing such individual practices should improve the potential for optimal training and, in turn, to optimised performance. The strategies, which are implemented, are therefore not just about performance enhancement on the day, but are demonstrated as useful and important for daily training regimes, which underpin performance.

From the scientific perspective, we still have much to learn regarding the responses of Paralympic athletes in stressful environments. The 2016 Games hosted by Rio de Janeiro will again present a range of environmental challenges and our athletes have to be well prepared. Although athletes with spinal cord injury are considered to be at greater risk of heat injury and have reported thermoregulation difficulties (4) there is little epidemiological data regarding incidence of heat injuries. There is a paucity of data on heat acclimatisation processes in those with SCI, although this is an issue currently under examination in our laboratory. The challenge for science is to provide rigorous testing in order to understand the physiological responses of Paralympic athletes whilst still being able to provide guidelines for applied settings. To this end, scientists and athlete support teams must work together to translate scientific results into sporting performance. Our understanding will certainly be enhanced by development of specific performance based protocols and individualised performance-enhancing strategies disseminated across sports.

References

1. Price MJ. Thermoregulation during exercise in individuals with spinal cord injuries. *Sports Medicine*, 2006;**36**(10):863-879.

2. Price MJ, Campbell IG. Effects of spinal cord lesion level upon thermoregulation during exercise in the heat. *Medicine and Science in Sports and Exercise*, 2003;**35**(7):1100 – 1107.

3. Webborn N, Price MJ, Castle PC, Goosey-Tolfrcy VL. The effects of two cooling strategies on thermoregulatory responses of tetraplegic athletes during repeated intermittent exercise in the heat. *Journal of Applied Physiology.* 2005;**98**(6):2101-7.

4. Martinez SF. Medical concerns among wheelchair road racers. *Physician and Sports Medicine.* 1988;**17**:63-68.

CHAPTER 11

Human Adaptation to Extreme Environments: the Limits of Personality

Benoit Bolmont[1] and Aurelie Collado[2].
[1] Université de Lorraine, LCOMS (Emotion-Action), EA 7306, Metz, F-57070, France
[2] UFR SciFA, Dept STAPS (Sport Sciences), Campus Bridoux, Avenue General Delestraint, Metz, F-57070, France

Acknowledgments
We thank institutions for the financial support (e.g., CNES – French Spatial Agency and Region Lorraine).

Vignette
There a plethora of possible causes proposed to explain for poor adaptation for athletes performing in extreme environments. Of these causes to poor adaptation, evidence suggests personality has an important role (1). For example, in conditions such as a simulated ascension of Everest or a parabolic flight, studies have demonstrated some individuals are more sensitive to environmental stressors (2,3), and re-appraised the situation by adopting a more realistic outlook on the size of the challenge. In addition, Abraini (4) reported adaptation difficulties in an extroverted diver, a finding that is consistent with previous research that shows extroverts find confined conditions or prolonged isolation in extreme environments difficult. In other specific extreme environments, authors have suggested that conscientiousness, agreeableness, and neuroticism play a role in the degree of adaptation (5). In polar experiments, a personality profile of being serious-minded, holding a self-sufficient attitude coupled with being able to cope with receiving less social support appeared advantageous (6).

In the present chapter, after presenting the main differences between extreme conditions, we consider interactions between personality and selected variables including social relations, human performances or affective disorders in the adaptation condition that could produce a consistent pattern of findings. In conclusion, we summarize the strategies aimed at helping athletes manage adaptation to extreme environments.

Discussion
Living and performing physical activity in an extreme environment is potentially stressful. As such, extreme environments could be a natural laboratory for investigating how people cope and adapt. Generally, living and performing intense exercise at extreme environments such as high altitude, deep diving, polar stations, or space flight associates with a number of intense physiological and psychological responses. Evidence has found that in a prolonged isolation and confinement experiments, factors in the social environment rather than the physical environment is the most potent source of stress (7). Examples of factors in the social environment include prolonged isolation, confinement, boredom, lack of privacy, reductions in the gratification, interpersonal relationships, etc.. Examples of factors in the physical environment include hostile natural conditions, physical dangers, hardships, night-daylight variations, lack or excess of sensory stimulations, hyperbaric, hypobaric, hypoxia or weightless conditions, and acclimatization. However, factors in the physical environment are the main source of stress in conditions such as parabolic flight characterized by on-going changes gravity or high altitude characterized by hypoxia (8, 9). Physical demands and social demands are multiple and influence people differently according to the extremity of the environment. For instance, Acute Mountain Sickness, a specific symptomatology at high altitude, characterized by a set of physiological, psychological, behavioural, cognitive and affective disorders due to hypoxic environmental conditions is related to both level of altitude and rate of ascent and vary between individuals (8,10). Whatever the features of extreme environments, given the rigors of physical and social life in these extreme environments, adaptation will be challenging and some people will adapt must better and much faster than will other people.

Evidence from studies suggests that the extent to which an individual adapts effectively depends on her or his personality traits. In Antarctic winter overs, studies have found that personality traits such as emotional stability play a role in the quality of interpersonal relations (11). Personality has also been found to influence performance in extreme conditions (12-14). Studies in polar winter overs or expeditioners have found that emotional stability, social compatibility, extraversion, assertiveness and neuroticism were individual characteristics related to effective adaptation and successful performance (15-18). In hypoxic situations, a high score on 'conscientious' personality trait scale was found to be related to poor performance in stimulus-response tasks (19). During a simulated ascension of Everest in a hypobaric chamber, Nicolas et al. suggested that the personality traits of *praxernia* (happening sensitivity) could play a major role in the occurrence of state-type anxiety responses. Similarly, Abraini et al. (4) synthesized studies that investigated the contribution of selected personality traits in the generation of anxiety in divers participating in experimental deep-dives with long-term confinement. Three divers from the sixteen participants demonstrated a clinically relevant level of anxiety during the course of the dives. Interestingly, personality patterns of these three divers show no identifiable differences from the remaining thirteen. They are characterised by low scores for self-control and emotional instability, which reflect incapacity to face and control external realties and happenings, i.e., to control and express tension in an appropriate manner. Thus, it would play a dominant role in the pattern of state-type anxiety (2,4,20,21). Similarly, evidence has suggested that diving-anxiety occurred in individuals who presented both a lack of personal adjustment and a considerable difficulty in establishing the personal rapport necessary for effective team participation (20, 21).

Conclusion

Evidence presented in the current chapter emphasises difficulties faced when attempting to draw a uniform synthesis. Extreme environments, experimental designs, conditions (actual vs. simulated) duration of exposure, and methods of assessment vary highly between studies, and not surprisingly, sample size tends to be small. All of these factors make it difficult to generalize findings. However, a common thread in the literature is the notion that personality is a key factor. Although it should be noted that physiological factors, sociodemographic factors, such as age, education and occupational status (5), and other psychological factors, such as coping strategies, adjustments also play a role in the adaptation processes.

A limited number of strategies could be used to identify individuals with limited adaptive capabilities. It is of course possible to use self-selection. It is proposed that each person estimates how he or she would cope with situational challenges and subsequently choose to volunteer or not. To provide people with a reasonable estimate of situational challenges, we suggest that people should go through some psychological preparation before the mission starts. This training could induce psychological preparation, such as stress inoculation training, or involve responses to environmental challenges and the individual should conduct some stress management training or perform some relaxation techniques.

A second option is consider how to select people who could be predicted to cope when performing in extreme environment is through preliminary psychiatric screening. Whilst psychiatric screening would provide useful information, it is important to recognise that many of the challenges are social in origin. Therefore, we suggest that it helps to investigate social factors such as interpersonal relationships with crewmembers, and the extent to which group members form a cohesive group. It should be noted that although extreme conditions may have a salutogenic or health enhancing effect (1,22), in general, extreme environments remain hostile and dangerous for human beings, a finding that suggests that every precaution must be taken before the mission. It is suggested that future research should consider the reciprocal relationship between individual variables and situational variables. Lane et al. (23) proposed a guide for using mood to assess the effects of adverse conditions so that interventions could be implemented to reduce the intensity of unwanted mood states. Future research should develop guidelines for researchers and practitioners alike.

References

1. Palinkas LA, Gunderson EK, Johnson JC, Holland AW. Behavior and performance on long-duration spaceflights: evidence from analogue environments. *Aviat Space Environ Med.* 2000 Sep;**71**(9 Suppl):A29-36.

2. Nicolas M, Thullier-Lestienne F, Bouquet C, Gardette B, Gortan C, Richalet J, Abraini JH. A study of mood changes and personality during a 31-day period of chronic hypoxia in a hypobaric chamber (Everest-Comex 97). *Psychol Rep.* 2000;**86**(1):119-26.

3. Collado A, Willmann M, Caillet G, Hainaut J, Bolmont B. Adaptation to parabolic flights: implications of personality and mood states (preliminary results). Paper presented at the 11th ESA Life Sciences Symposium Life in Space for Life on Earth, Trieste, Italy. 2010.

4. Abraini J, Ansseau M, Bisson T, de Mendoza J, Therme P. Personality patterns of anxiety during occupational deep dives with long-term confinement in hyperbaric chamber. *J Clin Psychol.* 1998 Oct;**54**(6):825-30.

5. Palinkas LA, Suedfeld P. Psychological effects of polar expeditions. *Lancet.* 2008 Jan 12;**371**(9607):153-163.

6. Butcher JN, Ryan M. Personality stability and adjustment to an extreme environment. *J Appl Psychol.* 1974;**59**(1):107-109.

7. Mullin CS. Some psychological aspects of isolated Antarctic living. *Am J Psychiatry.* 1960 Oct;**117**:323-325.

8. Bahrke M, Shukitt-Hale B. Effects of altitude on mood, behaviour and cognitive functioning. A review. *Sports Med.* 1993;**16**(2):97-125.

9. Schneider S, Brummer V, Gobel S, Carnahan H, Dubrowski A, Struder H. Parabolic flight experience is related to increased release of stress hormones. *Eur J Appl Physiol.* 2007 Jun;**100**(3):301-8.

10. Shukitt B, Banderet L. Mood states at 1600 and 4300 meters terrestrial altitude. *Aviat Space Environ Med.* 1988 Jun;**59**(6):530-2.

11. Sandal GM, Endresen IM, Vaernes R, Ursin H. Personality and coping strategies during submarine missions. *Mil Psychol.* 1999;**11**(4):381-404.

12. Mallis MM, DeRoshia CW. Circadian rhythms, sleep, and performance in space. *Aviat Space Environ Med.* 2005 Jun;**76**(6 Suppl):B94-107.

13. Morphew M, MacLaren S. Blaha suggests need for future research on the effects of isolation and confinement. *Hum Perf Extrem Environ.* 1997 Jun;**2**(1):52-53.

14. Sandal GM, Leon GR, Palinkas L. Human challenges in polar and space environments. *Rev Environ Sci Biotechnol.* 2006 Jun;**5**(2-3):281-296.

15. Sandal GM, Vaernes R, Bergan T, Warncke M, Ursin H. Psychological reactions during polar expeditions and isolation in hyperbaric chambers. *Aviat Space Environ Med.* 1996 Mar;**67**(3):227-234.

16. Rosnet E, Le Scanff C, Sagal MS. How self-image and personality influence performance in an isolated environment. *Environ Behav.* 2000 Jan;**32**(1):18-31.

17. Palinkas LA, Gunderson EK, Holland AW, Miller C, Johnson JC. Predictors of behavior and performance in extreme environments: the Antarctic space analogue program. *Aviat Space Environ Med.* 2000 Jun;**71**(6):619-625.

18. Steel GD, Suedfeld P, Peri A, Palinkas LA. People in high latitudes: the "Big Five" personality characteristics of the circumpolar sojourner. *Environ Behav.* 1997;**29**(3):324-347.

19. Bolmont B, Bouquet C, Thullier F. Relationships of personality traits with performance in reaction time, psychomotor ability, and mental efficiency during a 31-day simulated climb of Mount Everest in a hypobaric chamber'. *Percept Mot Skills.* 2001 Jun;**92**(3 Pt 2):1022-30.

20. Bugat R. Stress et plongée profonde. *Neuro Psy.* 1989;**2**:93-102.

21. O'Reilly J. Hana kai ii: a 17-day dry saturation dive at 18.6 ATA. VI: Cognitive performance, reaction time, and personality changes. *Undersea Biomed Res.* 1977 Sep;**4**(3).297-305.

22. Rivolier J, Goldsmith R, Lugg D, Taylor A. Man in the Antartic. London: Taylor & Francis; 1988.

23. Lane A, Terry P, Stevens M, Barney S, Dinsdale S. Mood responses to athletic performance in extreme environments. *J Sports Sci.* 2004 Oct;**22**(10):886-97.

Commentary: *Neil Weston*, University of Portsmouth, UK.

Research examining human adaptation to activities performed in extreme environments is not only intuitively interesting (1), but it could also provide essential knowledge that could ultimately help performers survive difficult circumstances (2). Whilst previous extreme environment research has examined a variety of parameters including mood (3), cognitive performance (4), and stress and coping (5), the present chapter focuses on the important role of personality in influencing human adaptation to such environments. Bolmont and Collado provide an overview of personality literature in this area, clearly and correctly emphasizing the idiosyncratic nature of human adaptation to performance in extreme environments. Associated with this theme, is the acknowledgement that each extreme environment contains unique physical and social demands that characterize and influence adaptation to that setting. For instance, a solo sailor in the southern ocean might experience significant sleep deprivation, physical exhaustion, mental fatigue and loneliness, whereas an individual performing as part of a polar expedition team may experience interpersonal difficulties, boredom, sensory deprivation and lack of privacy.

The literature is both blessed and hindered by the multitude of varying research designs, participant populations and differing dynamic environments, which have been studied thus making it difficult to determine any definitive conclusions. That being said, Bolmont and Collado do attempt to highlight the role of some personality variables in facilitating or hindering adaptation to extreme environments. Furthermore, the authors outline the impact that certain personality traits can have upon cognitive processing and affective adaptation to extreme environment stressors.

Evidence suggests that despite the lack of concrete findings regarding which personality dispositions facilitate human adaptation to extreme environments, organizations are using personality inventories to determine an individual's suitability for selection. Clearly further research is needed to ensure that using personality profiling to determine participant selection for performance in extreme environments is founded upon a valid and reliable scientific evidence base. Bolmont and Collado also emphasize the fact that a performer's personality can impact upon cognitive and affective responses in extreme environments and hence more research is needed to examine the interplay of these variables. This is of course dependent upon performers being willing to participate in "during-performance" research which is understandably not always top of their priority list when attempting to cope with the various challenging stressors. Hence, researchers should endeavour to find a careful balance between the practicalities of the extreme environment, the performer's needs and the employment of scientifically rigorous research designs in order to examine the link between personality traits and during-performance cognitive/affective responses.

Finally, greater dissemination of applied practitioner experiences using personality profiling in extreme environment activities is needed to elucidate the application of more traditional experimental research into real-world settings. Such publications will help to uncover key personality inventories that can be employed, the pitfalls to avoid and key strategies or methodological approaches to facilitate a successful use of personality profiling within extreme environments.

References

1. Suedfeld P. Extreme and unusual environments. In D. Stokols & I. Altman (Eds.), *Handbook of environmental psychology* (pp. 863-887). New York: Wiley; 1987.
2. Weston NJV. Learning to cope in extreme environments: Solo ocean endurance sailing. In Thatcher J, Jones M, Lavallee D (Eds.), *Coping and Emotion in Sport.* London, UK: Routledge; In Press.
3. Lane AM, Terry PC, Stevens MJ, Barney S, Dinsdale SL. Mood responses to athletic performance in extreme environments. *J of Sports Sci.* 2004;**22**:886-897.
4. Maruff P, Snyder P, McStephen M, Collie A, Darby D. Cognitive deterioration associated with an expedition in an extreme desert environment. *Brit J of Sports Med,* 2006;**40**:556-560.
5. Weston NJV, Thelwell RC, Bond S, Hutchings N. Stress and coping in single handed around the world ocean sailing. *J Appl Sport Psychol* 2009;**21**:460-474.

CHAPTER 12

Preparing for a Multiple Gold Medal Challenge- A Case Study of a Paracyclist

Gary Brickley
School of Sport and Service Management, University of Brighton, Eastbourne, UK.

Vignette

Disability cycling or Paracycling has grown exponentially over recent years in participation as well as performance standards. At the Paralympics, paracyclists with a range of conditions including visually impairment, amputees, spinal cord injury, and cerebral palsy compete in events on the track and road.

The present case study concerns Darren Kenny who has agreed to share this information on his preparation for Beijing 2008. Darren was strong junior cyclist who damaged his head during a crash in the Junior Tour of Ireland at the age of 18 years old. The crash and two further injuries resulted in a brain injury that led to him acquiring cerebral palsy (CP) which affects his left side. After approximately 15 years of not exercising, Darren returned to competition and competed in the Paralympics in Athens in 2004. In 2008 in Beijing, he attempted to win multiple gold medals, with a goal of winning the most medals by any individual in cycling at the Paralympic Games. Darren is 1.78m tall and has a body mass of 69.7kg and a power output at $\dot{V}O_2$ max of 407W.

To become a multiple medallist requires overcoming numerous challenges in preparation for the event. As a coach / exercise physiologist to Darren we first considered the energy requirements for each event. The events range from 250m sprint as part of the Team Sprint (approximately 21s) to the 1h 30 min road race (in the 36°C of Beijing). Also he had the 3000m pursuit (3 min 36 s), time trial (40 min) and the 1 kilometre time trial (1 min 10s) events to consider.

Over five years, physiological assessments were carried out in the field and the laboratory to assess his conditioning. Training was designed to ensure peak condition for major championships and for qualification. Given the diversity of the events and, consequently, varied physiological demand, there was a requirement to work on peak power production for events such as the Team Sprint and kilometre and the sprint at the end of the road race, and endurance capacity for the road events and 3km pursuit. Other important factors that influence performance were also considered, such as the ability to recover, coping with the heat and optimizing his equipment given his disability.

In the laboratory we would typically carry out an incremental test to determine lactate threshold on an SRM 'Ergometer' (Schoberer Rad. Messtechnik, Fuchsend, Germany) followed by a 20W/min ramp protocol to determine maximum aerobic power. A 6s all out test is used to determine peak power. The lactate threshold data is used to determine training zones and the ramp test determines aerobic fitness, whereas the 6 second maximal and all out test is useful for determining performance for starts on the track and sprints in the road race.

To ensure Darren was in the best possible physical condition to take on his multiple medal challenge, his laboratory records have been tracked over the years and are shown below in Figure 1. As Figure 1 indicates, he appeared to peak for the 2004 and 2008 Games. However, it is also interesting to note there are low performances lower than 340W during periods where he has been injured or taken a break from training. Having knowledge of Darren's physiological capabilities is important in planning performance and training as well as ascertaining his boundaries for performance.

In the field, critical power determination was used using a 1-minute, 3-minute, 5-minute and 10-minute all out self-paced effort. Critical power was derived from these 4 efforts as described elsewhere (1). Average power over the 3-minute test has also been found to be an excellent correlate of maximum aerobic power.

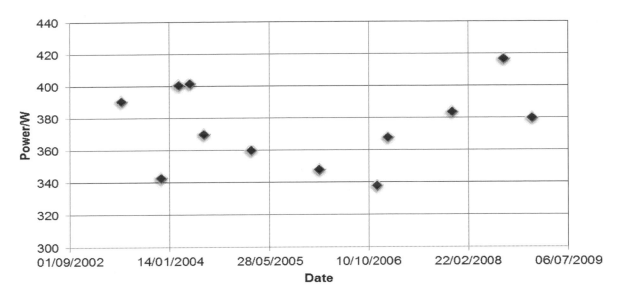

Figure 1: Variation in power (watts) of an athlete over 2 Paralympic cycles

There were three primary challenges in Beijing for Darren; First, preparation to peak for the multiple events; secondly, performance and finally recovery. We therefore put together 'targeted indicators'. For example, in the 3km pursuit we set a goal of 3 min 36 sec and achieved it but needed to 'break down' the event as below.

1st kilometre
Some of the kilometre specific work will complement this area;
- Strength – Over geared standing ½ and ¼, Standing 500's
- Power – Flying 500's, Race Gear standing laps
- Acceleration – Rolling Seated laps, rolling out of the saddle ½ laps
- Strength endurance – Standing 750's, Rolling Kilo's

2nd kilometre
- Motor-paced 4-8 laps over-speed;
- Cadence tolerance – Under-geared (115 rpm) flying 6-8 laps
- Short term Muscular endurance – Standing Kilos, Standing 6 laps

3rd Kilometre
Roadwork will complement and improve endurance;
- Lactate tolerance – Broken 2km over-speed.
- Cadence endurance – Under-geared broken and rolling 3km
- Technique tolerance – Rolling 4km efforts race cadence

Darren won his heat for the 3km pursuit comfortably before setting a world record in the final. Based upon his laboratory work, we estimated the average power needed to achieve this and specific progressive training was used to realize this target during the build up to the Games. The kilometre event typically lasts around 70 seconds for Paralympic athletes, which appears to be 50-60% aerobic. Therefore, it was important that Darren did not compromise his aerobic training in preparation for the kilometre time trial. We used 250-500m track work to ensure adequate preparation for the kilometre event in addition to specific work aimed at improving aerobic conditioning. This also complemented the 250m – 21 second lap for the team sprint. After 3 events Darren had 3 gold medals.

In addition, given Darren has a weakness in one side of his body, due to his cerebral palsy; we carried out torque analysis (see figure 2)(2).

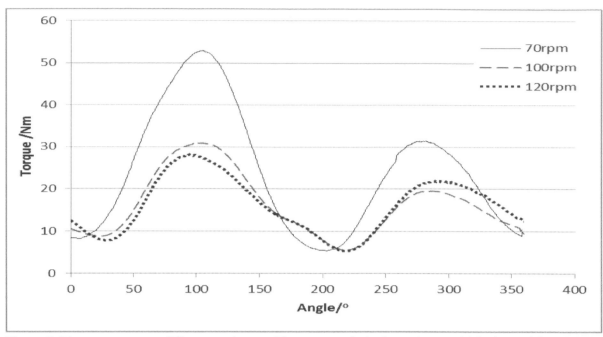

Figure 2. Torque curves at different cadences. The zero angle is the point at which the pedals are most vertical (2).

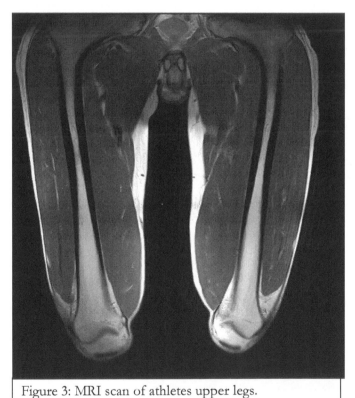

Figure 3: MRI scan of athletes upper legs.

From this, it is clear there are differences in torque production at the lower cadences. This was used to determine gearing during racing and training but also to use an elliptical chain ring to maximise power production. We were able to have an MRI scan carried out on Darren's legs to determine if there were any differences in muscle mass as a result of CP which may impact on function and performance but none were apparent (see figure 3).

The road race was more challenging, especially with regard to ambient heat and the competition would be more intense. In preparation for the race we travelled to the course 4 months before the event to film and collect power data on the course.

Darren opted to use a Camelbak 'drinking system' (Camelbak Products, Petaluma, CA) device on the time trial, which was useful for ensuring he could take on some fluid during the race but this interfered with his race helmet and may have reduced his time trial performance by 10s which resulted in a silver medal in that event. His pacing, based upon his laboratory data was excellent. The day after the time trial, Darren returned to win the road race where his strong aerobic ability ensured he stayed with the leader until the sprint finish and his 1200W peak power resulted in him winning the sprint. In recovery, we ensured that Darren minimized unnecessary ambient movement as his cerebral palsy condition makes walking very tiring. We ensured pre and post cooling was carried out during the road events and always ensured that a high protein / low carbohydrate solution was consumed after the race.

In summary, preparing an athlete for multiple events is very challenging. The athlete must be finely tuned with specific training to cover every event. Recovery and the ability to be optimally prepared for each event are crucial for successful performance.

References

1. Brickley G, Doust J, Williams CA. Physiological resposnes during exercise to exhaustion at critical power . *Eur J of App Physio*, 2002;**88**:146-151
2. Brickley G, Gregson HA Case study of a cerebral palsy cyclist using torque analysis. *Int Jof Sports Scie and Coach*, 2011;**6**(2):269-272

Commentary: *Louis Passfield*, University of Kent, UK.

The case study of a Paralympic cyclist provides insights into the demands of high-performance Paralympic cycling. In addition, this unique data is fascinating to evaluate and reflect upon. Darren's maximum power output of 417 W is typical of that of a highly trained cyclist. By calculating his power-to-weight ratio (6.0 W·kg^{-1}) you can make an approximate comparison with other track riders who might typically reach 7.4 W·kg^{-1} in peak form. In other words, Darren's maximum power output is approximately 20% lower than I've estimated for his able-bodied, world-record holding counterpart. It is notable however, that his asymmetry is much more marked than this, with his left leg producing up to 40% less torque than his right. Whilst the asymmetry associated with cerebral palsy is unsurprising, it raises important questions for training and competition. An obvious implication for training, is whether one should seek to "even out" the asymmetry, perhaps by targeting the affected side with single-legged training? Whilst not explicit on this point it seems that Gary Brickley and Darren have chosen instead to focus on developing strengths rather than addressing weaknesses. By sensibly making much of Darren's training drills to be competition specific they have effectively managed to overload both legs during cycling.

The data from the MRI scan reinforces the view that this seems to have been an effective strategy. This is because the scan suggests that the lower torque Darren generates with his left leg is neither the cause nor the consequence of a reduced muscle mass. Darren's asymmetry makes his peak power output of 1200 W particularly impressive. This figure is typical of elite road riders and suggests that he would be capable of holding his own in any bunch sprint, except against the best sprint specialists. It would be interesting to evaluate the degree of asymmetry in Darren's torque profile present during maximal sprinting activity in addition to the endurance-paced efforts shown in Figure 2. If his left leg torque is compromised during such maximal efforts to a similar extent as at lower intensities it will have a corresponding impact on his sprint performance. At lower intensities the opposite leg can compensate for some asymmetry. At maximum effort a lower effective torque from one leg will directly reduce the power output and cycling speed attained. Training for both endurance and peak power is also problematic. First, Darren is forced to divide his training time between two markedly different types of activity. Secondly, there is some evidence to suggest that the adaptations that result from endurance training may also reduce maximum sprint power output. Overall though, it is clear from the reported measurements and Darren's race results that his training programme has indeed been successful in developing both endurance and sprint capability to a very high level.

The International Paralympic Committee have a thankless task trying to promote relevant competition by disability category. Darren's marked asymmetry leaves one wondering whether such factors can and should be taken into account; and ultimately if appropriate competition categories for Paralympians are just too complex to be possible. The final paragraph of the case study highlights the careful holistic approach to preparation typical of many elite competitors, with the importance of equipment, pacing and nutrition all being carefully considered. However, the salutary tale of the drinking system demonstrates that even the prepared and motivated athlete can be wiser after the event.

CHAPTER 13

Live high (2000m), Train low (1050m) in Short-track Speed-skating.

Charles Pedlar

Centre for Health, Applied Sport and Exercise Science, St Mary's University, Twickenham, UK

Vignette

In 2004, prior to the 2006 Winter Olympic Games in Torino, Italy, the Great Britain short-track speedskating team (n = 9 male, 2 female; mean ± *s.d*; age: 20.2 ± 2.2 years; body mass: 70.7 ± 7.2 kg) explored the potential to conduct a training camp at moderate altitude with the aim of enhancing sea level performance and a view to including altitude training in their Olympic preparation. The period of 'altitude' exposure included 8 days of training at a moderate intensity in a normobaric hypoxic environment (F_IO_2 = 0.143, 75 mins·day^{-1}, 8 days) at the national training centre in Nottingham, UK, followed by 14 days of living in Bormio, Italy at 2000m (total period 22 days). Whilst some training was undertaken at 2000m and above during this period, daily skating training was completed at 1050m (~ 90 mins·day^{-1} variable intensity). The skaters lived in close proximity, with 2 or 3 skaters to each hotel room. Data collected were examined as a group and on an individual basis by a team of sports science and coaching staff in order to judge the overall benefits of altitude training for this squad prior to the 2006 Winter Olympic Games. The head coach to plan the allocation of funding for training camps, at least 2 years prior to the Games, used results.

Discussion

The duration of Olympic short-track speed skating events (longest event: 5000m relay: ~ 7 mins, shortest event 500m sprint: ~ 45 seconds) require skaters to be both aerobically and anaerobically efficient (1). Furthermore, the nature of the sport, particularly the tight bends skated at high velocity, requires skaters to hold a position which demands a high level of postural strength in a 'sitting' position. This 'sitting' position can lead to reduced asymmetric oxygen delivery and extraction, and a higher accumulation of blood lactate, possibly due to a reduced blood flow to the legs (2).

A number of studies have demonstrated that training at moderate altitude (1,500 – 3,000m above sea level) can improve oxygen uptake, transport, and delivery at sea level (3). This occurs through a number of mechanisms, potentially the most important being an augmented erythropoiesis, resulting in increased total haemoglobin mass (tHbmass)(4). Training at moderate altitude reduces maximum oxygen uptake and therefore oxygen flux during high intensity exercise and this can result in some detraining. A 'Live-high, train-low' strategy combats this problem because high-intensity training is conducted at a lower altitude where a higher oxygen flux can be achieved (5). Moreover, short-track speed skating performance is influenced by moderate altitude, due to the effect of the reduced air density on ice conditions (6) and on aerodynamic drag during skating (7) resulting in faster skating velocities. Anecdotally, skaters above 1000m altitudes commonly report 'faster ice'. Thus, living and training at altitude may be beneficial for, and is widely used by speed skaters, either for improved training quality or for the potential benefits of acclimatisation to altitude upon aerobic and anaerobic energy production (5, 8).

Amongst the range of deleterious effects of short altitude sojourns, two main candidates are poor sleep and inadequate recovery. Residence at moderate altitude or simulated altitude has been shown to cause sleep disruption in some individuals (9,10,11), therefore, there is a possibility that recovery from training may be hindered at altitude while acclimatisation takes place. It is important, therefore, to balance the potential gains from altitude training (5) against the potential lack of adequate recovery due to sleep loss. Furthermore, sleep deprivation reduces daytime function (12) and increases the potential for illness (13) which we seen collectively can further reduce the overall quality of training and the overall risk of maladaptation to altitude training (14).

Speed skating race. Photo courtesy of Charles Pedlar.

Intervention

In order to assess the impact of a 2-week altitude training camp, a range of positive outcome indicators were identified guided by theory and research that was available in the literature in 2004. These indicators included haematological parameters and measures of speed skating performance. In order to assess the potentially negative effects of altitude training, sleep quality and iron status were monitored. Routine measurements were taken daily during the training camp including urine osmolality, mood (Profile of Mood States), bodyweight, resting heart rate, blood pressure, peripheral oxygen saturation, and blood urea. These variables were examined collectively in order to assess signs of overreaching and illness (15) as well as altitude acclimatization, and were used by the coaches to monitor health and wellness and to make small adjustments to the training programme of the squad. Sub-maximal exercise tests were also regularly performed to monitor adjustment to the new training environment and individually adjust training programmes, particularly during the first week at altitude.

Findings

The majority of athletes suffered upper respiratory tract infections to varying degrees of severity during the altitude camp. These may have been caused by the additional stress of living at altitude, but it is equally plausible that the cause was non-altitude related, e.g. close living quarters, exposure to infectious agents during air travel etc.

Sleep: Sleep was monitored at baseline (7 nights prior to ascent to altitude) until 7 days after descent to sea level using wristwatch actigraphy, a non-intensive data collection method that measures movement and does not impact on the athlete in any way. In order to assess the group effects of altitude, data were grouped into four 7-day periods as follows: pre-camp at sea level (SL1), week 1 at altitude (ALT 1), week 2 at altitude (ALT 2) and post-camp (SL2). Disruption to sleep at altitude was observed on an individual level, for example, suppressed sleep efficiency (Figure 1) and lengthened sleep latency (Figure 2). Some of the sleep disruption may be attributable to other factors in the local environment, for example roommates; or an increased training load, however, it is likely that altitude played a role.

Maximal performance: Nine days before (RT1) and eight days after the altitude training camp (RT2), a maximal skating test was performed on the ice to measure skating performance (time to complete the test), maximum blood lactate concentration, and maximum heart rate. The test was discontinuous, involving 7 x 1.5 laps (166.68 m) of a standard short track; each skated at a maximal intensity. Between each 1.5 lap repetition, the skater cruised around the track, building up the speed for the next repetition over a period of 1 minute. This is a test regularly used by the team to measure performance, known as the 'Relay Test' and as such the skaters were well habituated to the demands of the test and the protocol. Performance time significantly improved ($P<0.05$) pre- to post-camp in all cases from 108.5 ± 3.3 seconds to 106.0 ± 2.9 seconds accompanied by trends towards lower mean heart rate and lower peak blood lactate concentrations. See Figure 3 for the spread of the performance improvement between individuals.

Blood: Venous blood samples were drawn and analysed locally. Evidence of accelerated erythropoiesis was observed with a trend towards higher circulating erythropoietin at altitude, which appeared to be at the cost of iron stores evidenced by a trend towards lower ferritin at altitude (see Figure 4). No measurements of tHbmass were taken which would have provided more definitive evidence of enhanced haemoglobin content and therefore aerobic capacity; however, over the relatively short altitude residence significant increases in tHbmass are unlikely to have occurred.

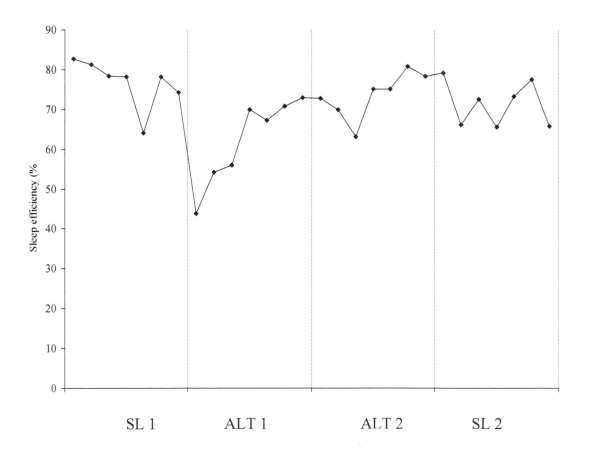

Figure 1: Sleep efficiency in one speed skater, where the lowest values were recorded during nights 1-3 at altitude, representing the worst quality sleep. The dashed vertical lines demark the 4 1-week periods SL1 = pre-camp week at sea level, ALT1 = week 1 at 2000m, ALT2 = week 2 at 2000m, SL2 = post-camp week at sea level.

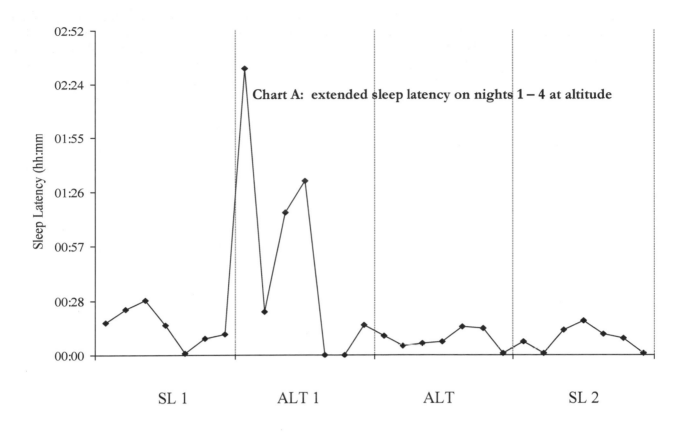

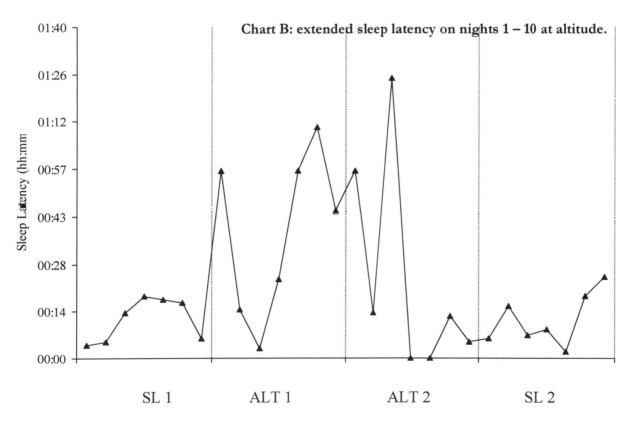

Figure 2: The impact of the altitude training camp upon sleep latency in two speed skaters. Chart A shows an extended sleep latency on nights 1 – 4 at altitude. Chart B shows extended sleep latency on nights 1 – 10 at altitude. The dashed vertical lines demark the 4 weeks SL1 = pre-camp week at sea level, ALT1 = week 1 at 2000m, ALT2 = week 2 at 2000m, SL2 = post-camp week at sea level.

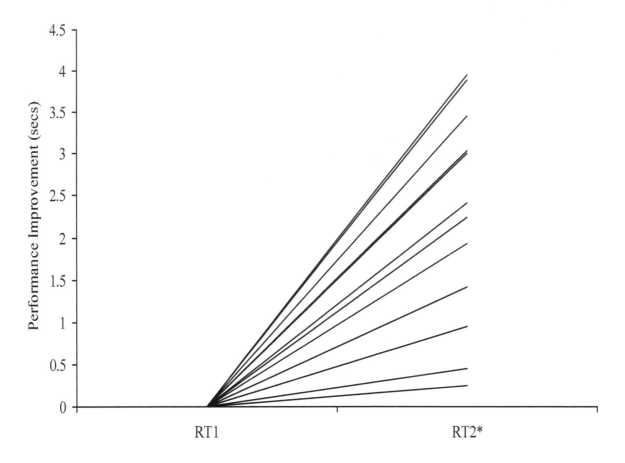

Figure 3: Absolute performance improvement in a maximal relay test comparing a test undertaken eight days before the altitude training camp (RT1) and eight days after the altitude training camp (RT2). * denotes a significant difference between tests ($P<0.05$).

Conclusions

The performance improvements observed in the present case study are similar to those observed in the published altitude training studies however; cause and effect cannot be stated here due to a lack of a control group (a common short-fall of altitude training studies).

Ultimately, the main outcome was that the coach made the decision not to employ a live high, train low strategy in the ensuing training and competition phases because he felt that whilst the intervention was not deemed to be a failure, the weight of positive evidence was not strong enough to adopt a longer term altitude strategy. The disruption caused to the training programme, the occurrence of upper respiratory tract infections and the negative impact on sleep quality coupled with a relatively small improvement in performance, did not justify the inclusion of altitude training in the immediate future.

This case study demonstrates the integration of objective physiological, performance and sleep behaviour data with coaching into an elite national squad, in order to plan a pre-Olympic strategy.

*Please note that the individual opinions summarised in this case study are those recalled by the author and may not be the opinions of GB Speed skating staff

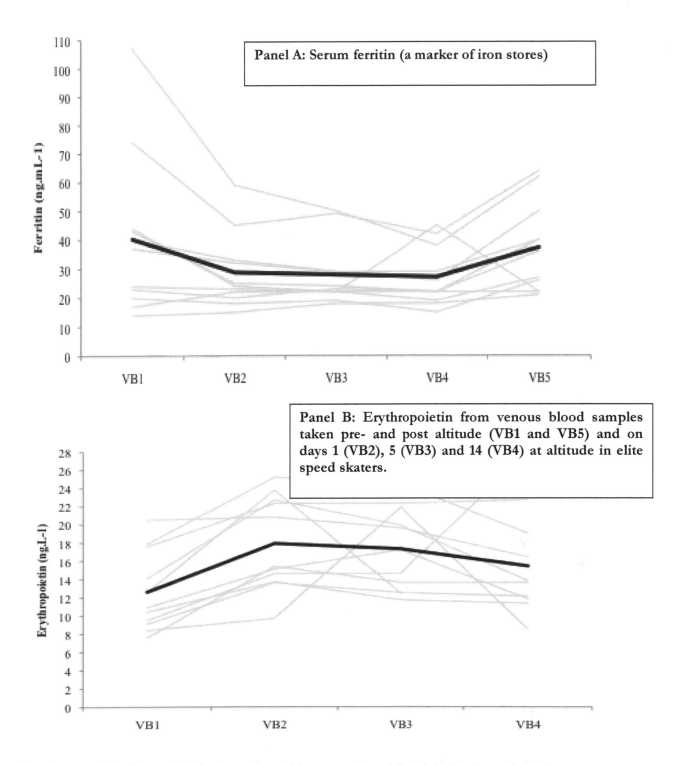

Figure 4: Panel A: Serum ferritin (a marker of iron stores) and Panel B: Erythropoietin from venous blood samples taken pre- and post altitude (VB1 and VB5) and on days 1 (VB2), 5 (VB3) and 14 (VB4) at altitude in elite speed skaters. The mean response (bold line) and individual responses (faded lines) are displayed.

References

1. de Koning JJ, Foster C, Bobbert MF, Hettinga F, Lampen J. Aerobic and anaerobic kinetics during speedskating competition. *Medicine and Science in Sports and Exercise.* 2003;**35**(5):s364.
2. Hesford CM, Laing SJ, Cardinale M and Cooper CE. Asymmetry of quadriceps muscle oxygenation during elite short-track speed skating. *Medicine and Science in Sports and Exercise.* 2012;44(3):501-8

3. Rusko HK, Tikkanen HO, and Peltonen JE. Altitude and endurance training. *Journal of Sports Sciences.* 2004;22:928–945

4. Gore CJ, Sharpe K, Garvican-Lewis LA, Saunders PU, Humberstone CE, Robertson EY, Wachsmuth NB, Clark SA, McLean BD, Friedmann-Bette B, Neya M, Pottgiesser T, Schumacher YO and Schmidt WF. Altitude training and haemoglobin mass from the optimised carbon monoxide rebreathing method determined by a meta-analysis. *British Journal of Sports Medicine.* 2013; 47(S1):i31-9

5. Levine BD and Stray-Gundersen J. "Living high-training low": effect of moderate-altitude acclimatisation with low-altitude training on performance. *Journal of Applied Physiology.* 1997;83:102–112.

6. Wilber RL. *Altitude training and athletic performance.* Human Kinetics, Leeds, UK.

7. Rundell KW (1996) Effects of drafting during short track speed skating. *Medicine and Science in Sports and Exercise.* 2004;28(6):765–771.

8. Lundby C, Millet GP, Calbet JA, Bartsch P and Subudhi AW. Does 'altitude training' increase exercise performance in elite athletes? *British Journal of Sports Medicine,* 2012;46:792-5.

9. Kinsman TA, Hahn AG, Gore CJ, Wilsmore BR, Martin DT, Chow C. Respiratory events and periodic breathing in cyclists sleeping at 2,650m simulated altitude. *Journal of Applied Physiology.* 2002;92:2114–2118.

10. Yaron M, Lindgren K, Halbower A C, Weissberg M, Reite M, Niermeyer S. Sleep disturbance after rapid ascent to moderate altitude among infants and preverbal young children. *High Altitude Medicine and Biology.* 2004;5(3):314–320.

11. Pedlar C, Whyte G, Emegbo S, Stanley N, Hindmarch I, Godfrey R. Acute sleep responses in a normobaric hypoxic tent. *Medicine and Science in Sports and Exercise.* 2005;37(6):1075–1079.

12. Leger D, Metlaine A, Choudat D. Insomnia and sleep disruption: relevance for athletic performance. *Clinics in Sports Medicine.* 2005;24:269–285.

13. Mazzeo RS. Altitude, exercise and immune function. *Exercise Immunology Reviews.* 2005;11:6-16.

14. Bailey DM, Davies B, Romer L, Castell L, Newsholme E, Gandy G. Implications of moderate altitude training for sea-level endurance in elite distance runners. *European Journal of Applied Physiology and Occupational Physiology.* 1998;78(4):360–368.

15. Meeusen R, Duclos M, Foster C, Fry A, Gleeson M, Nieman D, Raglin J, Rietjens G, Steinacker J, Urhausen A. Prevention, diagnosis, and treatment of the overtraining syndrome: joint consensus statement of the European College of Sports Science and the American College of Sports Medicine. *Medicine and Science in Sports and Exercise.* 2013;45(1):186-205.

Commentary: *Andy Lane* and Tracey Devonport, University of Wolverhampton UK.

The case study presents a scenario that can occur many times in sport science support with elite athletes in that multiple factors influence performance and these factors fall outside of the skillset of a single sports scientist. In the case provided above, altitude is a factor associated with changes in a number of physiological states. Pedlar describes a package that proved effective in terms of eliciting desired physiological adaptation. However, undesired consequences were also observed in terms of upper respiratory tract infections and sleep disturbance. In commenting upon this case study, we began considering a number of other variables that could be considered in evaluating the pro's and con's of attitude training (and living) that would influence the process. One such factor is an athlete's self-confidence might deteriorate. When an individual exercises at altitude it feels harder for the reasons described in the case study above, and so the possibility that the athlete confidence deteriorates represents a real possibility. Self-confidence has been found to predict performance and inversely relates to unpleasant and unwanted emotions (1). It is the link between self-confidence and motivated behaviour that could be especially important here. Confidence predicts the persistence and intensity of effort in the face of adversity whereby people high in self-confidence persist for longer and harder (1). Therefore, if confidence deteriorates, and is followed by a reduction in the intensity of training, then the goal to improve performance will not be achieved.

At altitude, a physiologist might expect a reduction in the intensity and duration of sessions. The athlete will be informed that exercise at altitude should feel harder, but how will he or she know how to gauge whether he/she is working hard enough? Speed will be slower and indices such as heart rate and breathing depth and frequency will be higher. The athlete will be acutely aware of this sensory feedback. A psychologist would aim to encourage athletes in his/her or her care to consider re-appraising symptoms of fatigue, and importantly, encourage ways of interpreting high heart rate scores as indicative of training intensity regardless of the likely fact that the speed achieved is slower. It is worth reflecting on the fact that worrying thoughts have been found to affect sleep and so clearly if an athlete believes he/she is not making progress at a training camp this will exacerbate the deleterious effects on sleep.

A suggestion to help an athlete prepare for an altitude training camp is to anticipate the likely stressors and have a solution prepared for this problem. Athletes, in particular, high-performance athletes, are acutely aware of the important role of sleep in the recovery process, and the consequences of upper respiratory tract infections for training outputs. An individual anticipating these as potential consequences of altitude training may experience negative thoughts towards altitude training. Should such thoughts remain unmanaged, then the potential consequences for confidence, training effort, engagement and interpretation of training outcomes may be undesired.

Research has found that a strategy called implementation intentions has been effective in coping with stress in a number of different areas including sport and health. Implementation intentions are formed as an "if-then" plan (2). The "if" part is the problem, for example, "if I notice my heart rate is unusually high for a 500m sprint" or "If I notice I feel out of breathe easily at altitude". The "then" aspect is the solution to the problem. Therefore, with regards to the present case study, the "then" aspect could be "I will tell myself not to worry about skating slower as the physiological adaptations are still taking place and these are positive". The athlete has to put the if-then component together so if I notice my heart rate is unusually high when running for a 500m sprint, then I will say to myself, not to worry about running more slowly as physiological adaptations are still taking place and these are positive". A benefit of if-then planning is its simplicity. The nature of an if-then plan is that the solution has been primed and so when the problem presents itself, the solution is the first thought that comes to mind. Furthermore, the simplicity of this type of intervention means that physiologists working in the field could utilise it. For example, it could be used to support athletes following instructions intended to accelerate adaptation to training at altitude.

References

1. Bandura, A. (1997). *Self-efficacy: The exercise of control.* New York: W. H. Freeman.
2. Gollwitzer PM, Sheeran P. Implementation intentions and goal achievement: A meta-analysis of effects and processes. *Adv in Exp Soc Psych.* 2006;**38**:69-119.

CHAPTER 14

Use of Iron Supplements in Young Rowers

Jeni J. Pearce
Performance Nutrition Lead, High Performance Sport New Zealand, Auckland, New Zealand.

Vignette

Two 17 year old club rowers (one male {MR}, one female {FR}) presented with symptoms of poor recovery from training sessions, fatigue, weakness, stable body mass, breathlessness and pale colour as reported by the coach. Body mass and body composition values were fairly typical and considered normal for individuals of this age and in this sport (Table 1). Blood data collected 12 months previously indicated values were within the normal range for age. Since this time the MR's height increased 5.5cm and body mass 4kg (predominately lean mass and bone). The FR grew 2cm and increased 3kg in body mass (prior body composition skinfold data was unavailable) and reported light and short duration (2 days) menses. Seven day dietary records including training sessions were obtained from both athletes and analysed (Table 1).

There was additional information that was provided from the MR's food diary. This indicated high intakes of semi skimmed milk as a beverage (2+L daily), providing a significant contribution to energy, protein and carbohydrate intake, no dietary supplement usage, with the majority of the iron consumed from non haem sources (red meat consumption was less than once a week). The MR's family consumed a high fibre (bran based cereals baked items and granary breads), heart healthy focused diet due to the father's high cholesterol and the mother's concerns for weight control. Hydration status was satisfactory with regular use of water and well-formulated sports drinks before, during and after training and races.

In terms of FR, additional data from the FR's dietary analyses revealed the daily use of a multivitamin and mineral complex (providing 14mg iron and 60mg vitamin C), an additional vitamin C supplement (providing 500mg vitamin C) and red meat consumed three times a week with a moderate fibre intake. A high protein diet (Atkins style) had been followed recently and ceased due to poor performances at training sessions. The fatigue persisted despite the reintroduction of carbohydrates, although intakes appeared to be below the recommended levels based on training sessions (Table 1). Both rowers were referred for routine bloods including a full blood count, ferritin and general medical review.

Discussion

Iron insufficiency is arguably the most common nutritional concern for athletes, coaches and support staff. Iron supplementation is reported among a range of sports from swimming (30%) to professional cycling (89%) (1,2). In non-anaemic athletes with low iron stores iron supplementation has been reported to lead to improvement (3, 4) in performance. Some athletes and coaches view supplementation as a preventative measure or providing an ergogenic benefit.

Seventy percent of iron in the body is found in haemoglobin, 3% in myoglobin with the remainder as ferritin (the major storage form). Reduced performance capacity and maximal aerobic power (impaired cellular respiration) has been reported with iron deficiency anaemia, especially in endurance athletes (particularly females and vegetarians) and when haemoglobin (Hb) levels are within physiological ranges (5).

Adequate iron status may be more important in athletes because, in addition to O_2 transport, iron is required, for a significant number of enzymes and cytochromes linked to energy metabolism and production, and also has roles in immunity and for cognitive function (6, 7). Additional iron is needed for the expansion of the blood volume occurring during growth (adolescence and pregnancy) and possibly in the early phases of training. Losses of iron occur with weight bearing athletes (runners) due to excessive sweating, gastrointestinal losses and bleeding, and haemolysis. Iron bound in the intestinal cells is lost when these cells are shed into the intestinal lumen at the end of the life cycle and removed via the faeces. Daily iron losses are relatively small in males (0.6-1mg/d via urine, faeces, sweat and desquamated cells) and half of that experienced by females due to menstruation (20-30mg per month) (8). The minimal requirement for iron absorption in men is 1 mg/d and 2-3mg/day for women and during adolescence (8). During periods of increased need (growth, blood loss and pregnancy) requirements increase to 4-5mg/d, returning to baseline once the need declines or no longer exists (8).

Haem iron, due to higher bioavailability (less affected by iron status, enhancers [vitamin C, meat binding facator] and inhibitors) provides two thirds of the 10-15% total daily dietary absorbable iron (5). Therefore, the type of iron available in the diet is important. Athletes are also encouraged to reduce the consumption of fibre (phytates), tea (tannins) and coffee (polyphenols) at mealtimes due to the inhibitory effect on iron absorption.

Female and adolescent athletes are particularly at risk for depleting iron stores, which, if left untreated, leads to anaemia, severely affecting the ability to train, recover, and compete. High carbohydrate diets, promoted to support high training loads, may hinder iron absorption due to the presence of inhibitors and unsuitable food combinations or dietary patterns reducing intake and compromising absorption. Poorly balanced vegetarian diets (strict vegan, microbiotic diets and disordered eating), low protein intakes and energy deprivation all lower available iron intake. Female vegetarians may be advised to increase iron intake levels to twice those of non-vegetarians due to poor bioavailability and the impact of inhibitors (phytates, polyphenols, calcium and soy protein isolates) (7). Vegetarians are encouraged to consume suitable iron-rich foods (cooked dried beans, dried fruit, and tofu) and consider taking a supplement containing iron, especially if active or participating in endurance activities.

Identifying and monitoring athletes at risk early in the training season is key as a full recovery from depleted stores is slow (3+ months) (9, 10) and delays the training gains. A 4-8 week time frame could produce a 30-50% increase in ferritin levels with oral supplementation (10). For a rapid recovery, supplementation is necessary, as dietary intervention alone takes significantly longer. Iron supplementation may induce gastrointestinal discomfort (constipation, increased stool frequency, black stools, abdominal discomfort such as cramping) and in some cases diarrhoea, nausea and vomiting. Inhibitory effect on zinc and copper absorption are recognised disadvantages of supplementation (calcium supplementation also reduces iron absorption).

Increased risk for cardiovascular disease and cancer has been suggested with excess iron intakes (red meat) and high serum ferritin, although results remain inconclusive (9). Dietary iron intake is unlikely to cause iron overload due to the regulation of absorption via the gastrointestinal system. However, there is no mechanism to actively remove excess iron. The adverse health effects of mild to moderately elevated iron stores have not been well reported. Iron overload could occur with iron supplementation, repeated iron injections and genetic disorders such as haemochromatosis (a genetic disorder). Absorption of 2mg/d, regardless of body stores can occur with haemochromatosis (absorption further increases after phlebotomy – up to 10mg/d) (9). Haemochromatosis leads to tissue and organ damage, including an increased risk of mortality from liver cancer and heart disease, with symptoms of fatigue, weakness and breathlessness reported. Avoiding supplementation in the absence of documented iron insufficiency and deficiency is advised.

Differences in the prevalence of iron deficiency occur due to the range in values (varying regionally and from country to country) used to measure and describe deficiency and the upper limits. Training induced declines in Hb, further increase the confusion over the diagnosis of iron deficiency. Low or suboptimal haematocrit or Hb concentrations alone are not necessarily indicative of anaemia, especially in endurance athletes or athletes undergoing rapid increases in training loads. Haemoglobin is regularly used for identifying anaemia and is insufficient to detect early or mild iron deficiency (13). Ferritin reflects the iron stores (not the content of Hb, myoglobin or enzymes) with values raised in the presence of inflammation, infection, and dehydration and following several days' exhaustive exercise, such as marathon running (6, 7, 11). One possible explanation is hemodilution due to the training effect (previously labelled sports anaemia or pseudoanaemia) that does not respond to iron supplementation and is short lived.

Accompanying indicators of the need to review iron status include low serum transferrin, low transferrin saturation and increased total iron binding capacity. Red blood cells may be affected by the presence of thalassaemia (inherited disorder of Hb synthesis). New measures such as reticulocyte development and

Hb content (showing bone marrow stores) (16) and ratios involving transferrin and ferritin are being investigated to provide greater clarity. The transferrin receptor ferritin index is not affected by infection or inflammation and is relatively stable (11). Iron storage is reflected in the serum ferritin value with the serum transferrin receptor providing the functional iron portion resulting in the ratio (11). A combination of measures may be required to indicate the difference between iron deficiency (depletion) and iron deficiency anaemia. It is important not to request tests in the dehydrated state or post hard endurance training sessions.

In body weight based sports, where energy intakes are compromised, impact nutrient density and available dietary-iron, place these athletes at greater risk. Athletes in these sports (distance running, ballet, gymnastics, diving, boxing, ice skating, martial arts, wrestling, weight-lifting and synchronised swimming), vegetarians and those with disordered eating, should receive regular dietary assessments to ensure sufficient dietary highly bio-available iron is provided. Iron supplements should be consumed only after a nutritional assessment by a qualified sports dietician/performance nutritionist and under medical supervision. Despite the limitations serum ferritin remains a viable choice for early detection of depleted iron stores until other measures such as transferrin receptors are validated. Of interest is the early evidence of increasing adiposity in women, lowering iron absorption and responses to fortification (14) although results appear equivocal (15).

Conclusions

The MR blood results reported Hb 11.7g/dL, haematocrit 0.40 L/L and serum ferritin 15µg/L. He was placed on ferrous sulphate 200mg twice-daily meals for 3 months. The athlete was advised of potential changes in stools and reflux (if consumed without food). Advice to consume the iron supplement with vitamin C rich meals, fruit juice and fruit (to enhance uptake) was provided, and consumption with high calcium dairy food and sports foods (protein shakes) was discouraged. At a family meeting the concerns of following overly 'healthy' regimes for young athletes in heavy training was discussed with the suggested introduction red meat three times a week, use of iron fortified cereals, wider selections in seafood and including meats closer to the bone (chicken legs and thighs) to raise iron intake. High fibre cereal intake (containing phytates) was reduced and replaced with lower fibre options. The provision of a wider variety of non-haem and haem iron will reduce the effect of inhibitors on iron absorption. A key recommendation was to consume mixed meals (containing meats as enhancers and vegetables containing vitamin C to improve non haem iron uptake) enhancing the total absorption of iron. The need to provide long-term dietary changes to prevent a recurrence of low iron status was reinforced. Intramuscular iron injections were not considered an appropriate option.

The majority of the blood results for FR demonstrated normal values, with the exception of an abnormally high ferritin (428ug/L) and a low total iron binding capacity. The athlete was referred for genetic testing for heredity haemochromatosis (later results proved positive). The athlete received counselling along with family members and was scheduled for periodic phlebotomy sessions. Dietary advice focussed on low iron foods, removal of all vitamin and mineral supplements containing iron and vitamin C, avoidance of iron fortified cereals and sports foods with added nutrients, with a focus on the ability to eat fish, low iron seafood, inclusion of tea and coffee with meals and eating seasonal foods (berries) in moderation. Carbohydrate intake was increased to 5-6g/kg.b.wt to support training sessions and recovery strategies were fine-tuned (timing of carbohydrate and protein ingestion post exercise). Regular monitoring was encouraged, especially if training was reduced (resulting from injury and the off season) and if menstruation returned more regularly. Both athletes reported improvements in general well-being, improved race times at trials (including one personal best and one seasonal best result) and the coach noted improved attitudes and recovery at training.

Decisions to use iron supplementation should be taken on a case-by-case basis, reviewed with blood measures (repeated at 3 and 6 months) and discontinued when serum ferritin or serum transferrin receptors are normalised. This is especially important as clear evidence on agreed lower limits of ferritin status remain controversial. Maintaining dietary changes to ensure the athlete consumes a diet with high levels of bio available iron is essential to maintain healthy iron status and avoid a recurrence of low status. Athletes and the general population may develop iron deficiency, but rarely iron deficiency anaemia (13), with iron depletion without anaemia and low Hb more prevalent and potentially limiting performance in sport and daily life. Early diagnosis is important and the risk of over diagnoses and over treatment of iron deficiency in athletes requires a Sport Science and Sports Medicine team approach.

Table 1
Athlete Characteristics and Dietary Analysis

Gender, Age	Body composition	Actual daily energy intake	Daily Macronutrient intake
Male, 17 yrs	178cm 74kg BMI 23 S8 64mm	14.4 MJ (3450 kcal)	535g CHO (62% E, 7.2g/kg.b.wt) 148g protein (17% E, 2g/kg.b.wt) 80g fat (21% E, 1.1g/kg/b.wt) 35g fibre
Female, 17 yrs	170cm 67kg BMI 23.1 S8 74mm	11.7 MJ (2800 kcal)	326g CHO (46.6% E, 4.9g/kg.b.wt) 137g protein (19.4% E, 2g/kg.b.wt) 105g fat (34% E, 1.5g/kg/b.wt) 22g fibre

S8 = ISAK Sum 8 skinfold sites

BMI = body mass index

E = Energy

References
1. Baylis A, Cameron-Smith D, Burke L. Inadvertent doping through supplement use by athletes: assessment and management of risk in Australia. *Int J Sport Nutr Exec Metab*. 2001;**10**:356-83.
2. Deugnier Y, Lorealm O, Carree F et al. Increased body stores in elite road cyclists. *Med Sci Sports Exec*. 2002;**34**:878-80.
3. Friedman B, Weller H, Mairbaurl H, Bartsch P. Effects of iron depletion on blood volume and performance capacity in young athletes. *Med Sci Sports Exec*. 2001;**33**:741-746.
4. Hinton P, Giordano C, Brownlie I, Haas J. Iron Supplementation improves endurance after training in iron-depleted, nonanemic women. *J Appl Physiol*. 2000;**99**:1103-1111.
5. Malczewska J, Raczynski G and Stupnicki R. Iron status in female endurance athletes and in non athletes. *Int J Sport Nutr Exec Metab*. 2000;**10**:260-276.
6. Eichner, E. Minerals: Iron. In Nutrition in Sport. Maughan ed. Blackwell Sciences, Oxford; 2000
7. Deakin V. Iron depletion in athletes. Chapter 10 in Clinical Sports Nutrition. Burke L, Deakin V. ed 3rd Edition. MacGraw-Hill Australia Pty Ltd, NSW, Australia; 2006; pg263-312.
8. Wood R, Ronnenberg A. Iron. In Modern nutrition in health and disease. Shils M, Shike M, Ross C, Carballero B, Cousins R (ed). 10th edn, 2006. Lippincott Williams & Williams. Baltimore.
9. Nielsen P, Natchtigall D. Iron supplementation in athletes. Current recommendations. *Sports Med*. 1998;**26**:207-216.
10. Dawson B, Goodman D, Blee T, Claydon G, Peeling P, Beilby J, Prins A. Iron supplementation: Oral tablets verses intramuscular injection. *Int J Sport Nutri Exec Metab*. 2006;**16**:180-186.
11. Malczewska J, Szczepanska B, Stupnicki R Sendecki B. The assessment of frequency of iron deficiency in athletes from the transferrin receptor-ferritin index. *Int J Sport Nutr Exec Metab*. 2001;**11**:42-52.
12. Auominen P, Punnonen K, Rajamaki A, Irjala K. Serum transferrin receptor and transferrin receptor ferritin index identify healthy subjects with subclinical iron deficits. *Blood*. 1998; **8**:2934-2929.
13. Landahl G, Adolfsson P, Borjesson M, Mannheimer C, Rodjer S. Iron deficiency and anemia: A common problem in female elite soccer players. *Int J Sport Nutri Exec Metab*. 2005;**15**;689-694.

14. Zimmermann M, Zeder C, Muthayya S, Winichagoon P, Chaouki N, Aeberli I, Hurrell R. Adiposity in women and children from transition countries predicts decreased iron absorption, iron deficiency and reduced response to iron fortification. *Int J Obesity.* 2008;**23**:1098-1104.
15. Ausk K, Ioannou G. Is obesity associated with anaemia of chronic disease? A populations-based study. *Obesity.* 2008;**16**(10):2356-2361.

Commentary: *Dr Bruce Hamilton* , Medical Lead, High Performance Sport NZ / NZ Olympic Committee, New Zealand

This excellent case study highlights many issues practitioners face when dealing with high-level athletes and coaching staff. Fatigue related underperformance is a common presentation in athletes, with a large differential diagnosis, warranting careful history taking, and it is well recognised that coaches and athletes will expect a blood test as part of the early evaluation (1). While the academic merit of early blood testing when faced with this clinical presentation causes ongoing debate (2), this report highlights that evaluating iron status via blood testing can in fact be a rewarding process, albeit not always in the manner anticipated. Iron overload has very serious metabolic implications, and as illustrated herein may present in a manner consistent with iron deficiency. Routine self-treatment by athletes and/or coaches with iron supplements should always be viewed cautiously and monitored carefully, particularly in endurance athletes (3). Furthermore, while the routine screening of elite athletes with a comprehensive array of blood tests may be controversial, evaluation of iron stores with a simple ferritin level appears warranted, particularly given both the prevalence of iron deficiency in women and the beneficial effects of supplementation (4). As highlighted in this case, both the upper and lower limits of ferritin levels considered appropriate in athletes remain an area of uncertainty, and this can result in a variety of different clinical practices when dealing with elite athletes.

Two elements of this case warrant further highlighting in the management of high performance athletes. Firstly, the anecdotal belief of some athletes and coaches that supra-physiological non-pathological levels (that is, high "normal" versus low "normal" levels) of ferritin are beneficial to endurance performance, which despite being widely held amongst high-level coaches and athletes has little evidence base. The danger of a belief based on the principle that a little is good, a lot is therefore better, is highlighted by this case. However, the role of iron in endurance performance, over and above recognised haematological benefits, is an area requiring further scientific evaluation.

Secondly, the belief of some elite level coaches and athletes that intra-muscular or intra-venous injections of iron will have additional benefits over and above oral supplementation. Both techniques for the injection of iron carry significant potential side effects, with both medical and medico-legal consequences, and in the absence of a clear indication should not be considered for routine use.

References

1. Fallon K. Blood tests in tired athletes: expectations of athletes, coaches and sport science / sports medicine staff. *Br J Sports Med* 2001;**41**:41-44.
2. Fallon K. Clinical utility of blood tests in elite athletes with short term fatigue. *Br J Sports Med* 2006;**40**: 541-544.
3. Lippi G, Schena F, Franchini M, Salvagno G, Guidi G. Serum Ferritin as a Marker of Potential Biochemical Iron Overload in Athletes. *Clin J Sports Med* 2005;**15**:356-358.
4. Fallon K. Utility of Hematological and Iron-Related Screening in Elite Athletes. *Clin J Sports Med* 2004;**14**:145-152.

CHAPTER 15

Individualizing fuelling and hydration for an Olympic endurance athlete

Trent Stellingwerff
[1]Canadian Sport Institute, Pacific, Victoria, B.C., Canada

Acknowledgments

A special thank-you goes to the 5-time Canadian Olympic race-walker Tim Berrett for consenting to show his data (especially for providing his split times from the 50 m race walk) and discussing his nutritional and physiological preparations for 2008 Beijing Olympic Games.

Vignette

The 2008 Beijing Olympic Games presented an extreme physiological challenge for endurance athletes given an average August daily temperature of $31^{0}C$ and humidity of 75% (hum). In conjunction, the 50km race-walk is a metabolically challenging Olympic event with an energy expenditure of ~3600 kcals during the ~4hr race. The race-walker (43yrs, 180cm, 67kg) in this case-study was making his 5th Olympic appearance. Hydration and fuel testing was longitudinally conducted prior to and throughout the 10-day pre-Olympic training camp in Singapore, and implemented into the 50km race-walk final, and featured: daily body-weight (BW) measures, hydration tracking (via urine specific gravity) and sweat rate and fluid consumption in anticipated weather conditions (Fig. 1; Table 1). Race day nutrition/hydration was precisely measured since athletes compete on a closed 2km loop with aid stations (Table 2). During the 10-day camp, the athlete's daily weight (67.5±1.6kg) and hydration status (1.012±0.006 g/ml) were relatively stable. During several key training sessions in anticipated weather conditions, individual sweat rate was 1767±208 ml/hr and practiced carbohydrate (CHO) and fluid intake were 67±14 g CHO/hr and 947±276ml/hr, respectively (Table 1). The pre-Olympic camp allowed this athlete many of the comforts of his normal Western culture, a similar time-zone to Beijing, and the required 10 to 14 days needed to allow for optimal physiological heat and jet-lag acclimation. Race day presented hot and humid conditions (10AM $29^{0}C$, 55% hum), and the athlete's directly measured intakes from the aid stations during the race was, on average: 1200ml/hr of fluid, 74g/hr CHO, 1120 mg sodium/hr and 100mg caffeine/hr (Table 2). All his CHO, fluid, electrolyte and caffeine targets were met on race-day with minimal GI side effects and only a 2.6% BW loss.

Discussion

It is nearly impossible to identify a single factor that causes fatigue and limits performance during endurance events lasting longer then 90min (1). Potential causes of exhaustion during prolonged high-intensity exercise can include central nervous system fatigue, thermal stress, dehydration, and most certainly involves mismanaged energy intake compared to expenditure. The 50km race-walk is one of the most metabolically and energetically challenging events on the Olympic program, with total estimated energy expenditure (EE) of ~3600 kcals during the ~4hr race. This EE is actually significantly greater than for either the triathlon or the marathon events. In this case study, several key science-based physiological and nutritional interventions will be highlighted that were tested and implemented with a 50km race-walker prior to and during the 2008 Beijing Olympic Games. All of the nutritional and physiological interventions are relevant to all endurance events lasting longer than 90min.

Singapore was chosen as the pre-Olympic camp for this Canadian athlete, as it allowed for the comforts of a normal Western culture, similar time-zone and weather conditions as Beijing, and the required 10 to 14 days needed to allow for optimal physiological heat and jet-lag acclimation (2). This ~2 wk period of acclimation allowed for an ideal balance between optimizing the relatively acute performance enhancing thermal adaptations versus having an overly prolonged daily heat strain negatively affecting final training quality leading into the championship (3).

Singapore and Beijing featured different cultures and foods, another time zone and differing weather conditions than this athlete normally faced (2). Therefore, daily BW and hydration tracking was implemented throughout the camp (Fig. 1). This allowed for individual and continual assessment, so that if any drastic changes or problems were presented, they could be immediately addressed and circumvented before they became a significant concern. Hydration was tracked via measurements of urine specific gravity (USG) using a hand refractometer (Fig. 1). To highlight, there was one workout on August 10th (Fig. 1; Table 1), in which this athlete inadvertently under-consumed fluid in relation to the extreme weather conditions. This substantial under-consumption of fluids resulted in a ~6% decrease in BW, and a large increase in USG, indicating a very dehydrated state. The 2007 American College of

Sports Medicine Position Stand on fluid intake illustrates the need for making individualized fluid intake recommendations according to individual sweat rates (4). Since Singapore had similar weather to Beijing, it provided ideal conditions to assess individual sweat rates (Table 1). Between two workouts that occurred on Aug. 10th and 12th, performed in slightly different environmental conditions, we were able to calculate an individual sweat rate of between 1.7 to 2L/hr. Since greater than a ~3% BW loss can lead to endurance performance decreases (5), we could estimate the amount of fluid intake per hour needed to prevent a greater than 3% loss in BW in targeted competition weather conditions (Table 1).

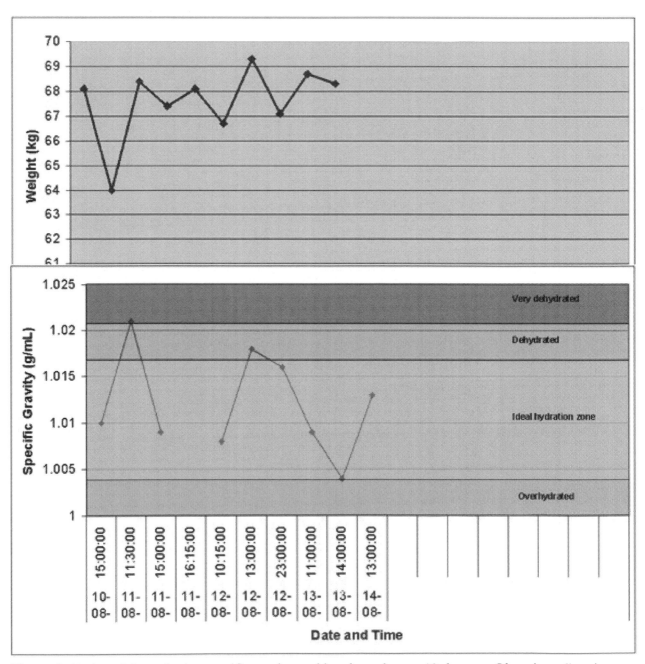

Figure 1. Body weight and urine specific gravity tracking throughout a 10-day pre-Olympic acclimation training camp for an Olympic 50km race-walker.

Table 1. Calculated sweat and fluid intake rates during acclimation camp and Beijing Olympic Final.

Parameters		Singapore Training Camp		Beijing
		Workout #1 Aug. 10th	Workout #2 Aug. 12th	Olympic Final Aug. 22nd
Weather	Temp (°C)	30°C	29°C	29°C
	Humidity (%)	78%	65%	55%
	Humidex	43	38	36
	Pre-weight (kg)	68.1 kg	68.1 kg	67.7 kg
	Post-weight (kg)	64.1 kg	66.7 kg	65.9 kg
	(% BW loss)	(5.9%)	(2.1%)	(2.6%)
Weight change		4.0 kg	1.4 kg	1.8 kg
		+	+	+
Total fluids consumed (L)		1.9 L	2.4 L	4.7 L
Total fluid turnover (L or kg)		5.9 L	3.8 L	6.5 L
		/	/	/
Exercise time (hr)		3 hrs	2.25 hrs	4.1 hrs
Sweat Rate (ml/hr) *		~2000 ml/hr	~1700 ml/hr	~1600 ml/hr
Actual Fluid Intake Rate (ml/hr)		~630ml/hr	~1070ml/hr	1140ml/hr
Calculated fluid intake rate to prevent >3% BW loss		~1270ml/hr	~760ml/hr	~1020ml/hr

* Urine losses not accounted for in sweat rate; 1 L water = 1 kg; BW - body weight; Temp - Temperature;

Beyond training and genetic status, the largest single determinant of ensuring optimal performance during prolonged endurance events in the heat is through the consumption of CHO energy, fluids and electrolytes. Data from several different labs has consistently shown that a CHO drink blend of glucose (GLU) and fructose (FRU) results in a ~30 to 50% higher oxidation rate as compared to GLU intake alone – the mechanism being that there are separate and distinct intestinal transporters for each GLU and FRU. Accordingly, several labs have confirmed that these GLU+FRU mixtures can lead to ~8% improvement in endurance performance (6, 7). Though, to achieve these high oxidation rates, large amounts of GLU and FRU need to be consumed (>1g/min), and this has recently been recommended to athletes when competing in endurance events >2hrs (8). However, large consumption of fluids and CHO during prolonged exercise can also result in adverse gastro-intestinal (GI) problems (9). Numerous anecdotal reports support the idea that an individual's GI tract can be trained and adapted to handle the intake of large amounts of CHO and fluid during exercise. In accordance, it has been suggested that the GI tolerance of large amounts of CHO intake can be optimized and adapted, although the ultimate amount appears to be very individual (8). Accordingly, recent evidence has also shown that the intake of high rates (>60g/hr) of GLU+FRU mixtures are very well tolerated in athletes during a field study featuring a 16-km running event (10). Contemporary data has also shown that CHO gels result in equivalent CHO oxidation as sports drinks (11). Therefore, a combination of sports drinks and gels were used by this athlete in training and during competition to reach goal CHO intake targets.

This individualized drink was the product of practical experiences drawn from a ~20-yr international race-walking career combined with the most recent scientific evidence and recommended targets as outlined above. A commercially available beverage provided the CHO base (2:1 ratio of GLU+FRU) of this individualized drink, which was augmented with extra electrolytes (Na+, K+, Mg+), caffeine and fluids to reach recommended and individualized targets. There is also sound evidence that caffeine intake enhances endurance performance in a small, but worthwhile amount (12). Current caffeine intake recommendations for athletes are ~3mg caffeine/kg BW consumed ~90 min prior to exercise. To ensure peak plasma caffeine levels throughout the 4 hr race, ~1.5mg/kg BW were dosed (~100mg) 60 min prior to exercise, and then again in the second, third and fourth hours, respectively (Table 2).

During the preceding competitions and months prior to the Games, this athlete became very accustomed to the CHO-based sports drink as outlined in Table 2. Impressively, this athlete adapted himself to be able to handle the intake of ~75g/hr of CHO (2:1 ratio of GLU+FRU) in conjunction ~1.2 L/hr of fluids (6% CHO solution) while walking 50km in just over 4hrs in extreme Olympic weather conditions. We targeted a large, but even distribution of CHO intake throughout the race, as time course data has suggested that immediate CHO intake upon the commencement of exercise will result in the greatest potential for muscle glycogen sparing (13). Taking scientific evidence in conjunction with experience, individual tolerances and flavour preferences resulted in a drink profile that this athlete had much experience and confidence with, and allowed for all his CHO, fluid, electrolyte and caffeine targets to be met with minimal GI side-effects and only a 2.6% BW loss on race day (Tables 1&2).

Table 2. Fluid, macronutrient and supplement intake for a 67kg male athlete during the 50km Olympic walk.

Drink characteristics	1st hour	2nd hour	3rd hour	4th hour	Average/hr	Targets
CHO (g):	75.6	70.8	64.8	85.4	74 g/hr	~60-90 g/hr
PRO (g):	11.34	10.62	9.72	9.06	10 g/hr	
Na+ (mg):	1190.7	1115.1	1020.6	1151.3	1119 mg/hr	~1 to 1.5g/hr
K+ (mg):	66.15	61.95	56.7	71.85	64 mg/hg	
Mg+ (mg):	113.4	106.2	97.2	90.6	102 mg/hr	
CAF pill (mg):	100	100	100	100	100 mg/hr	~6mg/kg over 4hrs
Total Fluids:	1245	1185	1110	1205	1200 ml/hr	~1 to 1.2L/hr

Conclusions

For elite athletes, nutrition and training interactions need to be carefully planned and monitored between coach, athlete and nutrition expert, and this comprehensive long-term approach needs to be considered for a range of different physiological factors. The featured interventions were individualized and periodized across the training and competition season, an approach starting to receive scientific attention (14). Testing and mimicking as many of the planned interventions as possible prior to the championship event (logistics, environment, nutrition, etc.), and considering and anticipating all possible scenarios, results in an athlete having the full psychological confidence in their unique and individualized approach. Ideally, this supports the athlete in realizing their best possible performance for their targeted championship event.

References

1. di Prampero PE. Factors limiting maximal performance in humans. *Eur J Appl Physiol* 2003;**90**:420-429.
2. Reilly T, Waterhouse J, Burke LM, Alonso JM. Nutrition for travel. *J Sports Sci* 2007;**25**:S125-S134.
3. Maughan R, Shirreffs S. Exercise in the heat: challenges and opportunities. *J Sports Sci* 2004;**22**:917-927.
4. Sawka MN, Burke LM, Eichner ER, Maughan RJ, Montain SJ, Stachenfeld NS. American College of Sports Medicine position stand. Exercise and fluid replacement. *Med Sci Sports Exerc* 2007;**39**:377-390.
5. Cheuvront SN, Carter III R, Sawka MN. Fluid balance and endurance exercise performance. *Curr Sports Med Rep* 2003;**2**:202-208.
6. Currell K, Jeukendrup AE. Superior endurance performance with ingestion of multiple transportable carbohydrates. *Med Sci Sports Exerc* 2008;**40**:275-281.
7. Triplett D, Doyle A, Rupp JC, Benardot D. An Isocaloric Glucose-Fructose Beverage's Effect on Simulated100-km Cycling Performance Compared With a Glucose-Only Beverage. *Int J Sport Nutr Exerc Metab* 2010;**20**:122-131.
8. Jeukendrup AE. Carbohydrate feeding during exercise. *European Journal of sport Science* 2008;**8**:77-86.
9. Brouns F, Saris WH, Rehrer NJ. Abdominal complaints and gastrointestinal function during long-lasting exercise. *Int J Sports Med* 1987;**8**:175-189.
10. Pfeiffer B, Cotterill A, Grathwohl D, Stellingwerff T, Jeukendrup A. The effect of carbohydrate gels on gastrointestinal tolerance during a 16km run. *Int J Sport Nutr Exerc Metab* 2009;**19**:485-503.

11. Pfeiffer B, Stellingwerff T, Zaltas E, Jeukendrup AE. Carbohydrate Oxidation from a Carbohydrate Gel Compared To a Drink during Exercise. *Med Sci Sports Exerc.* 2010;**42**(11):2038-2045.
12. Burke LM. Caffeine and sports performance *Appl Physiol Nutr Metab* 2008;**33**(6):1319-1334.
13. Stellingwerff T, Boon H, Gijsen AP, Stegen JH, Kuipers H, van Loon LJ. Carbohydrate supplementation during prolonged cycling exercise spares muscle glycogen but does not affect intramyocellular lipid use. *Pflugers Arch*, 2007;**454**(4):635-647.
14. Stellingwerff T, Boit MK, Res P. Nutritional strategies to optimize training and racing in middle-distance athletes. *J Sports Sci* 2007;**25**:S17-S28.

Commentary: *Andrew Drake.* Leeds Metropolitan University, UK.

Trent Stellingwerff's case study demonstrates how evidence-based practice helped an Olympic athlete meet race day nutritional targets preparing for the 50 km race walk in the 2008 Beijing Track & Field programme.

Competitive 50 km race walk pacing should benefit from nutritional input. Despite the successful intervention described in the case study, the athlete fatigued during competition as evidenced from 5 km split times. Even when identifying an optimal race intensity the athlete experienced a J-shaped positive pacing profile characteristic of high level 50 km race walkers, i.e. 5 km splits of ~23, ~23, ~24, ~23, ~24, ~25, ~25, ~26, ~28.5 and ~26.5 min (1). This profile was similar to the bulk of the Olympic games field with one exception: the Gold Medal Winner exhibited a negative pacing profile (2). Stellingwerff outlines potential physiological and regulatory processes causing fatigue and it appears there is certain inevitability to this in practice.

A common race walking economy curve resolved from laboratory measured oxygen uptake (VO_2) and race walking speed estimates the average VO_2 for a ~4 h 08 min 50 km race walk is ~51 ml/kg/min using the equation: VO_2 (ml/kg/min) = 5.2482v (km/h) – 12.334, where v = race walking speed (3). Certain assumptions are inherent computing a common race walking economy curve, i.e. the curve ignores individual differences in race walking economy. The average VO_2 ~51 ml/kg/min equates to ~4350 kcal (a little higher than identified in the intervention) and likely muscle and liver glycogen stores in the 67 kg athlete described are ~450-500 g (~1800-2000 kcal). Assuming glycogen stores are fully utilised there remains a shortfall - some offset by the 75-g/hr average carbohydrate (CHO) intake from the glucose/fructose drink blend and CHO gels ingested during the race. Moreover a CHO-loading diet in the days pre-competition is the likely practice; plus caffeine ingested in race could have impacted on several mechanisms, e.g. enabling calcium release to promote muscle contraction (4). Here it is important to acknowledge the need to practice and tolerate ingesting the drinks and gels described, which as stressed were individualised based on evidence and experience, with a further need to avoid dehydration in the thermally stressful environment.

The J-shaped positive pacing profile described suggests despite the support teams interventions it is difficult to maintain a uniform pace throughout a 50 km race walk particularly when the task is competing, as opposed to completing the distance. The demands of the event severely challenge energy intake and expenditure, primarily through CHO availability; therefore, it may also be advantageous to increase lipid availability in the form of free fatty acids (FFA). In this long term intervention specific preparation utilising physiological testing could aid the identification of an optimal training intensity and duration for an athlete seeking to improve FFA flux from capillary blood into muscle fibre so sparing CHO, i.e. increasing exercise capacity at a given race walking speed (5). Ad so returning to race day preparation, this case study demonstrates the value of sport science knowledge and skills successfully utilized in the applied setting.

References

1. Vernillo G, Agnello L, Drake A, Piacentini MF, La Torre A. An analysis of the pacing strategy during 50-km race walking events. *15th Annual Congress of the European College of Sport Science, Antalya, Turkey* 2010.
2. 2008 Olympic Games Athletics 50 Kilometres Race Walk Results *http://www.iaaf.org/OLY08/index.html* [26-04-2011]

3. Drake A, James R. Prediction of race walking performance via laboratory and field tests. *New Studies in Athletics* 2009:**23**(4):35-41.
4. Burke LM. Caffeine and sports performance *Appl Physiol Nutr Metab* 33, 2008.
5. Arcelli E. Marathon and 50 km walk race: physiology, diet and training. *New Studies in Athletics* 1996;**11**(4):51-58.

CHAPTER 16

Fuelling an Ironman World Champion

Asker Jeukendrup
Human Performance Lab, School of Sport and Exercise Sciences, University of Birmingham, Edgbaston, Birmingham B15 2TT, United Kingdom

Vignette

The Hawaii Ironman World Championship is not only challenging because of the sheer distances involved (3.8km swim, 180km bike and 42 km run) but also because of the environmental challenges with temperatures around 35°C, a humidity that is approaching 80% and fierce headwinds on the bike.

The triathlete in this case report is: female, 30 years of age, 57 kg and 176 cm. The athlete had turned professional in 2007 and qualified for the Ironman World Championships by winning Ironman Korea. Performance in Ironman distance triathlon is dependent on many factors but nutrition plays a crucial role (1). Often nutrition is called the 4th discipline of triathlon and many athletes who do not perform well attribute this to nutritional problems. Two inter-linked problems in endurance sports are gastro-intestinal distress (2) and running out of energy. This makes nutrition during such an event a fine 'balancing act' between consuming enough energy but making sure that absorption is adequate and no gastro-intestinal distress develops.

For the Hawaii Ironman event the most important nutritional issues identified included
1. Avoiding major gastro-intestinal (GI) distress,
2. Providing energy throughout the event and in a way that does not exacerbate GI distress in those susceptible to it, and
3. Maintaining fluid balance.

Managing GI distress

The athlete had a history of gastro-intestinal problems. These problems typically occurred later in the race and were predominantly lower GI tract problems. After analysing food intake in the days before it was noted that ingestion of fibre, as well as dairy products, was relatively high. Previous research has demonstrated that fibre intake is linked to an increased incidence of gastro-intestinal problems. The fibre intake of the pre-race meal (breakfast) as well as that in the days leading up to the race was reduced to a minimum. Similarly, dairy product intake was reduced to very low levels for 2 days.

Gastro-intestinal distress may also occur when more carbohydrate is ingested than can be emptied from the stomach and/or absorbed. Therefore the next strategy was to provide an energy source during exercise that is readily absorbed and oxidised resulting in minimal residual volume in the stomach and intestine. This will be discussed in the next section.

Over time the breakfast for this athlete has 'evolved' and so that ingested before her last few races has consisted of a bowl of cream of rice cereal made with water with a tablespoon of sunflower butter and a tablespoon of honey, half a banana and percolated coffee with lactose free milk. The energy content and composition of the meals is comparable.

Table 1: Intake ~3 hours before the race

Time	Food	Energy	Carbs	Fat	Protein	Fibre
4:00 AM	Large mug of tea with milk, 3 English muffins, 1 Banana, 2 slices of white bread with jam and cheese (cheddar), 1 slice of white bread with honey and cheese water	1364 kcal	221 g	34 g	56 g	7 g

Fuel stores and fuelling during the race

In order to start the race with optimal muscle glycogen stores the athlete consumed a high carbohydrate diet providing her with nearly 9 g/kg/day of carbohydrate (585g/day) (Table 2). The day before the race carbohydrate constituted 65% of her energy intake and fat intake was relatively low at 25%. Studies have demonstrated that well trained athletes will optimise their glycogen stores on such a carbohydrate intake even when they are in training (3). The athlete exercised for approximately 2 hours the day before the race but at a low intensity.

Liver glycogen must also have been high after the relatively large carbohydrate intake 3 hours before the race (221g; Table 1). So the athlete followed recommendations and so should have been able to start the race with optimal glycogen stores in both liver and muscle (4).

Table 2: Energy and macronutrient intake the day before the race

Intake day before	Energy	Carbs	Protein	Fat	Fibre
	3672 kcal	585 g	151 g	99 g	25 g

At 6:45AM the pro men and women started their race in Kona. The athlete had been sipping water till the start of the race but did not take anything else in the hour before. She took one gel (25 g carbohydrate, no caffeine) in the first transition (T1) before mounting the bike for a five-hour ride in the lava fields of the Big Island. Her fuel for the ride was waiting for her on the bike: two 750ml bottles of a highly concentrated carbohydrate solution (20 scoops of a carbohydrate-protein drink; vanilla flavour). At regular intervals (at least every 10 min) the athlete consumed a mouthful of this solution and washed it down with a sip of water from her water bottle attached to her handlebars. At every feed station she took a water bottle and used this to fill her handlebar mounted water bottle and also splashed water on her face and neck and to aid cooling. At the halfway point when she had moved into the top 10, Chrissie took a non-caffeinated gel and after 150 km when she had taken over the lead she had one more gel; this time the gel was caffeinated. Most of the energy during her ride, however, came from the bottles she had prepared before the race (see table 3).

With a 2 min lead she entered T2 where she consumed another caffeinated gel before starting the second fastest marathon of all times. During the marathon she would use the feed stations to grab sponges and ice to keep cool and water to wash down a gel. Her strategy was to have a gel every 25 min just before the feed station and wash this down with water. During the run she managed to take 7 gels (the first one of these coming out of transition). In the last hour she also sipped small amounts of non-carbonated cola drink at feed stations.

Overall her carbohydrate intake was calculated to be about 775 grams of carbohydrate during the bike and run. Over the entire race her average intake was 86 g/h, which is very close to the recommended 90 g/h (5). The carbohydrate source that was used is maltodextrin and fructose and because the combination of these carbohydrates results in a fast absorption it is possible to maintain such high intakes (6) without accumulation of the carbohydrate in the gut and consequent gastro-intestinal problems. Indeed the athlete experienced minimal problems during the race. These carbohydrate combinations have also been shown to result in higher exogenous carbohydrate oxidation rates (5, 7-9), better fluid delivery (10) and improved performance (11) compared with a "traditional" sports drink containing predominantly one type of carbohydrate.

It is worth noting that the athlete did not take additional salt tablets but relied on the extra sodium provided in the gels and some sodium present in her concentrated carbohydrate solutions.

In races to follow the protein was removed from the carbohydrate solution she consumed during the race and the athlete did not use protein in her drinks anymore. This is in line with the lack of convincing evidence for a benefit of added protein (12).

Table 3: Intake during the Ironman World Championship 2007 in Hawaii

Discipline	Time	Food	Energy	Carbs
Swim	0:58 h	-	-	-
T1	2:21 m	1 Gel	100 kcal	25 g
Bike	5:06 h	CHO	2000 kcal	500 g
		2 Gels	200 kcal	50 g
T2	2:03 m	1 Gel	100 kcal	25 g
Run	2:59 h	6 Gels	600 kcal	150 g
		Cola	100 kcal	~25 g
Overall	9:08		3100 kcal	775 g

Recently we wrote a letter to the editor (13) to indicate that one of the characteristics of a successful elite endurance athlete is a very high capacity to absorb carbohydrate. Clearly this athlete is a good example of this, being able to compete at very high intensity and still ingest and tolerate large amounts of carbohydrate. Genetics play an important role in this but the gut is extremely adaptable and trainable (7) and hence practice is recommended during training to establish optimal ingestion protocols for the individual.

Fluid intake

Fluid intake is always more difficult to measure and in this particular case we did not attempt to quantify fluid intake more accurately in this race but the important point to stress is that we uncoupled carbohydrate from fluid intake. The recommended carbohydrate intake is relatively fixed and independent of the environmental conditions. The fluid intake recommendations are not only highly individual but also dependent on the conditions. A fluid intake in a group of 52 Ironman triathletes at the 2009 Ironman World Championships was just under 800 ml/h (unpublished data) but there was a lot of variation (range: 425 ml/h -1520 ml/h). Body weight was not measured before and after the race so the fluid losses for the athlete cannot be estimated.

World Champion

The strategy employed was clearly successful as the athlete went on to win the race and was crowned World Champion. In years to follow she would repeat this in many other Ironman races every time making small improvements to the nutrition strategy, thereby reducing gastro-intestinal distress and improving the energy and fluid delivery. The drinks the athlete uses are now different (partly for sponsorship reasons), but the composition remains very similar.

Acknowledgement

A very big thank you goes to the 2007, 2008 and 2009 Ironman World Champion and world record holder for the Ironman distance Chrissie Wellington for consenting to discuss her nutritional strategy during the 2007 Ironman World Championship.

References

1. Jeukendrup AE, Jentjens RL, and Moseley L. Nutritional considerations in triathlon. *Sports Med.* 2005;**35**:163-181.
2. Brouns F and Beckers E. Is the gut an athletic organ? Digestion, absorption and exercise. *Sports Med.* 1993;**15**:242-257.
3. Coyle EF, Jeukendrup AE, Oseto MC, Hodgkinson BJ, and Zderic TW. Low-fat diet alters intramuscular substrates and reduces lipolysis and fat oxidation during exercise. *Am J Physiol Endocrinol Metab.* 2001;**280**:E391-E398.
4. Hargreaves M, Hawley JA, and Jeukendrup A. Pre-exercise carbohydrate and fat ingestion: effects on metabolism and performance. *J Sports Sci.* 2004;**22**:31-38.
5. Jeukendrup AE. Carbohydrate and exercise performance: The role of multiple transportable carbohydrates *Current Opinion in Clinical Nutrition and Metabolic Care.* In press, 2010.
6. Wallis GA, Rowlands DS, Shaw C, Jentjens RL, and Jeukendrup AE. Oxidation of combined ingestion of maltodextrins and fructose during exercise. *Med Sci Sports Exerc.* 2005;**37**:426-432.
7. Jeukendrup AE and McLaughlin J. Carbohydrate ingestion during exercise: effects on performance, training adaptations and trainability of the gut. *Annals of Nutrition and Metabolism.* In press, 2011.
8. Jeukendrup AE. Carbohydrate intake during exercise and performance. *Nutrition.* **20**:669-677, 2004.
9. Jeukendrup A. Carbohydrate feeding during exercise. *European Journal of Sport Science* **8**: 77-86, 2008.
10. Jeukendrup AE and Moseley L. Multiple transportable carbohydrates enhance gastric emptying and fluid delivery. *Scand J Med Sci Sports.* 2010;**20**:112-121.
11. Currell K and Jeukendrup A. Superior performance with ingestion of multiple transportable carbohydrates. *Med Sci Sports Exerc.* 2008;**In Press.**
12. Jeukendrup AE, Tipton KD, and Gibala MJ. Protein plus carbohydrate does not enhance 60-km time-trial performance. *Int J Sport Nutr Exerc Metab.* 2009;**19**:335-337; author reply 337-339.
13. Stellingwerff T and Jeukendrup AE. Authors reply to Viewpoint by Joyner et al. entitled "The Two-Hour Marathon: Who and When?" *J Appl Physiol.* 2010:**In press.**

Commentary *Mathew G. Wilson.* Qatar

Over the past couple of decades, prolonged endurance events (>8hrs) such as ultra-distance running and Ironman triathlon have become a global phenomenon. Whilst the Hawaii Ironman is considered the most iconic triathlon in the world, the Ironman brand currently has 26 official Ironman distance events, and 52 'half'-distance events. Needless to say, ultra-endurance events have become a commercial extravaganza that leave many athletes feeling bewildered in areas such as selecting the appropriate cycling equipment, balancing swim-bike-run training and of course, sports nutrition. It is correct to state that nutrition is the 4th discipline of triathlon, but from professional experience, it is the 4th most neglected discipline. There are 3 key elements to Prof Jeukendrup's case study to which I would like to draw the reader's attention:

1. The utilisation of food diaries in identifying a nutritional problem, in turn providing a racing solution,
2. The correct mathematical calculation of carbohydrate (CHO) required and;
3. Creating a nutritional race strategy based upon current scientific evidence not commercial branding.

The most important nutritional objective during an Ironman is to avoid GI distress. To avoid this, three variables must be taken into account;

1. The need to identify potential sources of GI distress through simple food diary analysis. In this particular case, high fibre and dairy intake was ascertained as a possible hindrance, allowing food modification to occur, helping to create a better racing platform,

2. If an athlete does not adapt their highly trainable digestive system to cope with a large volume of high energy (yet often sickly!) products under duress, the consequences aren't particularly pleasant,

3. Athletes will only know what nutritional products their gut through practice tolerates, and so experimentation should never be left until race day.

This case study also demonstrates well the calculation of the required number of grams of CHO required per hour based upon the estimated race time and tolerance to CHO concentration vs. gut absorption rates. By opting to use two pre-prepared bottles of CHO on the bike with only supplemental water to preserve fluid balance, this is an excellent example of effective race management to sustain energy for the run portion of the triathlon, with a trusted energy product ensuring the correct CHO dose per hour. It is interesting to note that the athlete drank water before the race and avoided CHO. Limited evidence exists for the impact of CHO ingestion prior to exercise causing a rebound hypoglycaemia (1) but arguably one in five individuals suffer with this so for those affected, timing will be even more crucial. Finally, the athlete did not use additional protein supplementation. The effective use of protein during exercise is still under debate, with some studies showing little to no performance improvements over CHO alone(2, 3). Indeed, the difficulties triathletes face with keeping abreast of the scientific evidence is not helped by commercial companies advertising the apparent benefits of a particular endurance supplement. Athletes should seek qualified guidance when confused (4).

In conclusion, exercising continuously for periods between 8 and 14 hours is not uncommon in Ironman. Prof Jeukendrup eloquently demonstrates the need to identify potential sources of GI distress and mathematically personalise sports nutrition, allowing the athlete to optimise their performance.

References

1. Jeukendrup AE, Killer SC. The myths surrounding pre-exercise carbohydrate feeding. *Ann Nutr Metab*. 2010;**57** Suppl 2:18-25.
2. Breen L, Tipton KD, Jeukendrup AE, No effect of carbohydrate-protein on cycling performance and indices of recovery. *Med Sci Sports Exerc*. 2010;**42**(6):1140-1148.
3. Jeukendrup AE, Tipton KD, Gibala MJ. Protein plus carbohydrate does not enhance 60-km time-trial performance. *Int J Sport Nutr Exerc Metab*. 2009;**9**(4):335-337; author reply 337-9.
4. Rodriguez NR, Di Marco NM, Langley S. American College of Sports Medicine position stand. Nutrition and athletic performance. *Med Sci Sports Exerc*. 2009;**41**(3):709-731.

CHAPTER 17

Hydration and Marathon Running Performance in the Heat

Louise M. Burke
Sports Nutrition, Australian Institute of Sport, Australia

Vignette

Following widespread publicity of adverse outcomes suffered by runners in a Big City Marathon, a coach consulted a sports dietician to advise his squad about hydration for an upcoming event. The squad included several competitive runners aiming to finish their first marathon in < 2 h 40 min, as well as a larger group with an estimated finishing time up to 2 h longer than this. The coach had previously provided a handout about hydration clipped from an old running magazine. He liked its simplicity, recommending a certain volume of fluid should be consumed at each aid station. However, a recent news item featured a sports scientist stating that such advice was wrong and dangerous, and that runners should only drink if they were thirsty. He was confused about what to tell his squad.

Discussion

Evaporation of sweat provides the main mechanism to dissipate the heat produced during exercise or absorbed from the environment. Rates of sweat loss during exercise vary according to factors such as the intensity of exercise, environmental conditions (heat, humidity, airflow) and the athlete's size, clothing and state of fitness/acclimatization. In various sports and exercise activities, sweat rates typically range from 0.5-2.0 L/hour (1), and can cause substantial losses of fluid and electrolytes. These losses are exacerbated if the athlete commences the session with a pre-existing fluid deficit; something that might occur if there has been inadequate opportunity to restore fluid balance following a prior bout of exercise or deliberate dehydration such as "weight making" for competition in a weight-division sport. Opportunities to offset these losses include pre-exercise hyper-hydration practices, and more importantly, hydration during the session. The development of guidelines to help athletes adopt appropriate drinking strategies before and during sport might appear to be a simple task. However, it has become a source of considerable discussion and controversy within sports science circles, as evidenced by the continual evolution of official guidelines (1, 2, 3) and heated criticism of such guidelines (4, 5).

Several aspects of early guidelines for fluid intake during exercise have been justifiably criticised or superseded by recent research findings. The first positions stands on this topic (6, 7) were focussed on distance running and provided recommendations that were overly prescriptive, unsympathetic to the individuality of needs, and conservative regarding the benefits of consuming carbohydrate as well as fluid during exercise of >1 h duration. In real life, the logistics or challenges of drinking during an exercise session will tend to vary between sports and individuals. Important issues include access to fluid (must the athlete carry their own supplies or are drinks available via aid stations or trainers/handlers?), opportunities to drink (are there breaks in play or does the athlete need to consume fluids, literally, "on the run"?), gastrointestinal comfort, and perception of fluid needs. These issues underpin the difficulty of developing prescriptive guidelines with sufficient flexibility to cover the demands of the myriad sports or exercise activities. A critique of seemingly sensible recommendations, even those targeting the apparently simple circumstances of a single sport, may find them to be inadequate and even harmful. For example, the suggestion distance runners should drink 100-200 ml of water at aid stations provided every 2-3 km in a race (6) could span the intake of 330 ml/h by a slow runner to 2 L/h by a fast runner (8). Such low intakes may range from inadequate to suitable for slower runners depending on actual sweat rates, but the high rate is likely to cause gastrointestinal discomfort as well as interfere with race performance by requiring the runner to slow down substantially at each aid station to obtain and consume the drink. Importantly, it is unnecessary to drink at rates that are in excess or, or even equal to, sweat rates.

Current guidelines by groups such as the American College of Sports Medicine (1) and the National Athletic Trainers Association (2) have tried to accommodate the individuality of needs and opportunities to hydrate during exercise across a range of sports. They recommend that each athlete develop an individualised fluid plan based on an appreciation of likely sweat rates, knowledge of opportunities to drink during the exercise session, and feedback/fine-tuning from experience. In addition, they acknowledge the potential additive benefits of refuelling during the event (9), via the consumption of carbohydrate-containing drinks or foods. Carbohydrate intakes of 30-60 g/h are typically recommended for activities involving > 60 min of moderate or intermittent high-intensity exercise (1, 2), although intakes of ~80 g/h may be optimal for ultra-endurance events of > 3-4 h (9). Periodic assessments of fluid balance and refuelling practices across an exercise session, particularly under conditions that simulate an event of interest, would allow each athlete to assess the suitability of their current drinking practices and judge whether these could benefit from further practice and manipulation (see Fig 1). However, even these recommendations have been subjected to the criticism of being unnecessary or complicated (5). Instead, it has been argued that athletes should simply drink according to their thirst (5).

Hydration targets are based on the goal of avoiding a fluid deficit that could impair performance or health. However, defining such a fluid deficit is a source of disagreement. There is substantial literature on the effects of different levels of fluid loss on exercise capacity or performance (1, 10, 11, 12). The outcomes vary according to many characteristics of the research design including the type of exercise, the environment in which it is undertaken, and whether the fluid deficit is incurred over the course of the session or was present at the start of exercise. Nevertheless, impairments can typically be detected in both physical (power outputs, work rates etc) and mental (skills/concentration etc) aspects of aerobic and intermittent high-intensity exercise undertaken in hot conditions when dehydration exceeds 2% of body mass (10, 12). The effects increase with the size of the fluid deficit but are smaller in cooler conditions, and it is equivocal whether such mild dehydration has a clear effect on single efforts involving power or strength, although strength endurance is impaired (11). Dehydration overlays on the thermal strain of exercise and may exacerbate problems in hot weather (1).

Accordingly, current guidelines recommend that personalised fluid plans for most sports or exercise activities aim to keep fluid deficits below ~ 2% of body mass, especially if the activities are undertaken in a hot environment (1, 2). Critics argue that the literature is almost entirely based on laboratory research in which the cooling effects of wind and air resistance (13) or the motivating effects of the sporting atmosphere is absent. As such the literature overestimates the true effect of dehydration on performance in the field. This must be counterbalanced by the consideration that traditional probability statistics combined with the small sample sizes typical of sports science research are only capable of detecting substantial reductions or differences in performance. Such analyses may fail to recognise performance changes that would be important to the outcomes of real-life sport, where events are decided by milliseconds and millimetres). Indeed, when the same exercise is undertaken in a hot environment with incrementally changing levels of fluid deficit from 0-4% body mass, there is a parallel change in thermoregulatory strain, cardiovascular drift and perception of effort (14). Intuitively, a similar subtle but clinically significant pattern may occur with endurance or performance, and some new studies report impairment of performance in field conditions with mild fluid deficits (15).

This is not, at least historically, a common occurrence in most sports < 2-3 hours where typical drinking practices replace ~ 50-70% of sweat losses/body weight losses (16). However, observational studies of sporting events over the past decade show that some individuals are overzealous with their interpretation of hydration guidelines and drink at a rate that substantially exceeds their sweat losses (17, 4, 5). Risk factors for this syndrome, which can lead to potentially fatal hyponatremia (low plasma sodium concentrations), include being female he observations of a range of patterns of fluid intakes/balance during competitive sports (20). An individualized cost-benefit analysis of fluid intake during a sport or event may identify practices, which optimize performance with minimal risk of gastrointestinal discomfort, time lost in drinking or risks of over-drinking. Athletes can assess the suitability of their background hydration practices (see Figure 1), with drinking to thirst or ad libitum intake providing a reasonable starting point. In various cases, however, there may be room to improve on these subjective dictates. For example, in sports where opportunities for fluid intake are limited, the athlete may need to drink at the available opportunities early in an event, "ahead of their thirst", to better pace total fluid intake over the session. In addition, as long as total fluid intake is not excessive, athletes might consume beverages such as sports drinks to meet carbohydrate refuelling targets rather than fluid replacement. Finally, we should remember that some individuals are unable to respond appropriately to thirst (or hunger) because we live in a food environment in which portions are continually upsized so that the ability to judge suitable amounts or volumes to consume is difficult.

STEPS
1. Weigh yourself before training, using reliable digital scales. This should be done wearing minimal clothing (underwear only if possible) and after going to the toilet
2. Weigh in again after training in the same clothing, and after towelling dry
3. Weigh your drink bottle before and after your workout to find out how much fluid you consumed. Or just estimate the amount of fluid consumed and convert ml of fluid into grams
4. Note the weight of any foods or sports products consumed during the session (e.g. gels, lollies, bars)

Extra steps for additional information or accuracy:
a. If you have to go to the toilet during the session, weigh in before and after you go or measure the volume of your urine
b. Estimate total urine losses during the session by going to the toilet after the post-session weigh in and reweighing again, or measuring the volume of your urine. Add this to the volume of urine produced at any mid-session toilet stops
c. Note pre-exercise hydration status. Urine characteristics before the session may tell if you are already dehydrated or in normal fluid balance. Strategies to over-hydrate should be also noted - the success of these may be shown via an increase in body mass.

CALCULATIONS (with illustration from a sample runner)
Your fluid intake (ml) = drink bottle before – drink bottle after (g)
e.g. 700 g – 100 g = 600 g or 600 ml (contained 6% carbohydrate sports drink)
Your urine losses (ml) = change in body mass due to toilet stops during and/or after the session (kg) x 1000
e.g. 60.25 – 60.00 = 0.25 kg = 250 ml
Your fluid deficit (ml) = body mass pre-session – body mass post-session (kg) x 1000. (Note: to measure total fluid deficit which includes sweat and urine losses, use value from the post-session weigh-in taken after the toilet stop)
e.g. 60.50 – 59.05 = 1.45 kg = 1450 ml
Your fluid deficit (% body weight) = (Fluid deficit [in kg] X 100)/ body mass (kg)
e.g. (100 x 1.45)/(60.50) = 2.4%
Total sweat losses over the session = Fluid deficit (g) + fluid intake (g) + food intake (g) – urine losses (g)
e.g. 1450 + 600 + 40 g (sports gel) – 250 = 1840 ml
Sweat rate over the session = sweat losses converted to ml per hour
e.g. Session lasted for 90 min: sweat rate = 1840 x 60/90 = 1000 ml or 1.0 L/hour
Note also the total carbohydrate content from sports drinks, gels, lollies or other foods
e.g. carbohydrate intake = sports gel (25 g) + 600 ml sports drink (6% carbohydrate = 36 g)
total carbohydrate = 61 g

Figure 1. Steps for calculating fluid balance and fuel intake over a session of exercise.

ADDITIONAL NOTES:
While this activity can help you to estimate the net fluid deficit incurred across a session of exercise, some interpretations are needed:
• During prolonged exercise at high workloads (> 2-3 h), some loss of mass from depletion of fuels (e.g. glycogen) can be expected, overestimating total sweat losses. In addition, water may be liberated from glycogen breakdown. Therefore, for marathons-ultra-endurance events, it may be sensible to adjust the estimation of the fluid deficit by subtracting 1-2 kg from
• Pre-event hydration status needs to be taken into account before assuming that the net fluid deficit over a session represents the true levels of dehydration. If the athlete has pre-existing dehydration, the total fluid deficit is underestimated. By contrast, if the athlete has hyper-hydrated prior to the session, the net fluid deficit will be overestimated

106

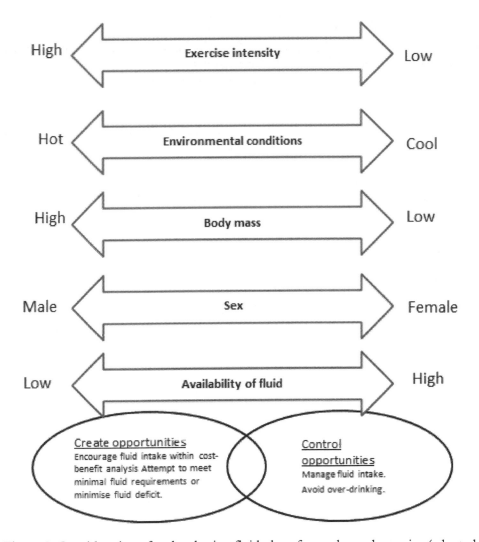

Figure 2. Considerations for developing fluid plans for prolonged exercise (adapted from (21))

References

1. American College of Sports Medicine, Sawka, MN, Burke LM, et al. American College of Sports Medicine position stand. Exercise and fluid replacement. *Med. Sci. Sports Exerc.* 2007;**39**:377-390.

2. Casa DJ, Armstrong LE, Hillman SK, et al. National athletic trainers' association position statement: fluid replacement for athletes. *J Athl Train.* 2000;**35**:212-24. Hew-Butler T, Verbalis JG and Noakes TD. Updated fluid recommendation: position statement from the International Marathon Medical Directors Association (IMMDA), *Clin J Sports Med.* 2006;**16**:283-292.

4. Noakes, TD and Speedy DB. Case proven: exercise associated hyponatraemia is due to overdrinking. So why did it take 20 years before the original evidence was accepted? *Brit J Sports Med* 2006;**40**:567-572.

5. Noakes TD and Speedy DB. Time for the American College of Sports Medicine to acknowledge that humans, like all other earthly creatures, do not need to be told how much to drink during exercise. *Brit J Sports Med.* 2007;**41**:109-111.

6. American College of Sports Medicine. Position stand of the American College of Sports Medicine: the prevention of thermal injuries during distance running, *Med. Sci. Sports Exerc.* 1987;**19**:529-533.

7. American College of Sports Medicine. Position stand: exercise and fluid replacement, *Med. Sci. Sports Exerc.* 1996;**28**:i-vii.

8. Coyle EF, Montain SJ. Carbohydrate and fluid ingestion during exercise: are there trade-offs? *Med. Sci. Sports Exerc.*1992;**24**:671-678.

9. Jeukendrup AE. Carbohydrate intake during exercise and performance. *Nutrition*. 2004;**20**:669-77. Cheuvront SN, Montain SJ, Sawka MN. Fluid replacement and performance during the marathon. *Sports Med*. 2007;**37**:353-7.

11. Judelson DA, Maresh CM, Anderson JM et al. Hydration and muscular performance: does fluid balance affect strength, power and high-intensity endurance? *Sports Med*. 2007;**37**:907-21.

12. Sawka MN, Pandolf KB. Effects of body water loss on physiological function and exercise performance, In: Gisolfi CV and Lamb DR (eds).*Perspectives in exercise science and sports medicine*. pp 1-38. Benchmark Press, Carmel, In. 1990.

13. Saunders AG, Dugas JP, Tucker R, Lambert MI, Noakes TD. The effects of different air velocities on heat storage and body temperature in humans cycling in a hot, humid environment. Acta Physiol Scand. 2005;**183**:241-55.

14. Montain SJ, Coyle EF. Influence of graded dehydration on hyperthermia and cardiovascular drift during exercise, *J Appl Physiol* 1992;**73**:1340-1350.

15. Casa DJ, Stearns RL, Lopez RM, et al. Influence of hydration on physiological function and performance during trail running in the heat. *J Athl Train*. 2010;**45**:147-56. Noakes TD, Adams BA, Myburgh KH et al. The danger of an inadequate water intake during prolonged exercise. *Eur J Appl Physiol*. 1988;**57**:210-219.

17. Almond CSD, Shin AY, Fortescue EB, et al. Hyponatremia among runners in the Boston marathon. *New Eng J Med*. 2005;**352**:1550-1556.

18. Montain SJ, Cheuvront SN, Sawka MN. Exercise associated hyponatremia: quantitative analysis to understand the aetiology, *Brit J Sports Med* 2006;**40**:98-106.

19. Hew-Butler T. Arginine vasopressin, fluid balance and exercise: is exercise-associated hyponatraemia a disorder of arginine vasopressin secretion? *Sports Med*. 2010;**40**:459-79.

20. Garth AK, Burke LM. What do athletes drink during competitive sports activities? *Sports Med*. 2013 **43**:539-564.

21. Burke L and Cox G. The Complete guide to Food for Sports Performance, 3rd edition. Allenand Unwin, Sydney.

Commentary: Susan M Shirreffs, Loughborough University, UK.

As highlighted in the case study, the point at which a change in hydration status (to a fluid deficit) will have an impact on performance has been researched and reported in the scientific literature. Although 2% is frequently highlighted as the body mass loss at which performance or capacity declines, there is individual variation about this mean. And as described in the case study discussion, the laboratory tests employed are probably not sensitive enough to identify the small changes in performance that are relevant to athletes. This perhaps represents the biggest hydration challenge to athletes; establishing how big a change in hydration status they can tolerate before their performance is negatively affected. If time permits in their schedule, athletes may benefit by purposely exercising at different degrees of dehydration, in environmental conditions that they expect they may need to perform in, and establish the level of dehydration at which they feel their performance may be affected. Once established, the athlete can then 'set their goal posts' for individualising their hydration strategy.

As described in the case study, having an individualised hydration strategy is generally desirable for athletes (1). Fortunately, with practice and fine-tuning after collection of some basic data, it should be achievable for all athletes. However, for some athletes, they may need to accept that they are going to become significantly dehydrated during exercise. For example a 70kg football (soccer) player with a sweat rate of 2.5 litres per hour is unlikely to be able to consume a water volume great enough over the course of a match to prevent significant dehydration developing. The rules of the game, his desire to remain in position on the field of play and possible gastrointestinal discomfort are likely to be reasons why this is the case. Therefore, recovery of any outstanding deficit after the end of the match will become a priority to ensure that by the time of the next training session or match, euhydration has been restored (3). This restoration can be determined by measuring body mass, if the athlete knows their normal, euhydrated body mass (2) in combination with another marker, for example; a measure of urine concentration as described in the case study.

References

1. American College of Sports Medicine, Sawka, MN, Burke LM, et al. American College of Sports Medicine position stand. Exercise and fluid replacement. *Med. Sci. Sports Exerc.* 2007;**39**:377-390

2. Cheuvront SN, Carter R 3rd, Montain SJ, Sawka MN. Daily body mass variability and stability in active men undergoing exercise-heat stress. Int. J. Sport Nutr. Exerc. Metab. 2004;**14**:532-540.

3. Shirreffs SM, Armstrong LE, Cheuvront SN. Fluid and electrolyte needs for preparation and recovery from training and competition. *J. Sports Sci.* 2004;**22**:57-63.

CHAPTER 18

Observations of Dietary Intake and Potential Nutritional Demands of a National Football Squad.

Justin Roberts.

School of Life and Medical Sciences, University of Hertfordshire, UK.

Vignette

In preparation for the 2010 South Africa World Cup, nutritional intake and diagnostic assessment was undertaken on the England National Squad (n=29). The purpose was to evaluate provisional nutritional intake/demands, with post assessment recommendations aiming to optimise dietary support. Urinary and plasma sampling for diagnostic assessment was undertaken on 29 players, with dietary intake recall diaries collected over a 4 day period for 16 players.

Average daily energy and carbohydrate intakes were 1998.6±412.2 kcal/d and 228.0±66.5 g/d respectively, both of which being below current recommendations for athletes. Average daily protein and fat intake was found to be 130.4 ±49.6 g/d and 66.0±28.8g/d respectively, both falling within current guidelines. Various recommendations including increased energy intake, balanced meal timings, portion size, glycemic loading and protein/carbohydrate ratios were addressed as part of the nutrition programme.

With challenges faced in assessing nutrition status or demand, the use of a relatively new model of functional analysis was carried out (NutrEval profile, Genova Diagnostics Ltd). Data indicated a number of areas where nutrient demand was elevated, despite sufficient recommended nutrient intake (RNI) levels. Whilst care should be taken to interpret functional tests, the combined use of dietary recall and diagnostic testing may be useful for providing nutritional support. Directed nutritional support should be undertaken several times throughout the season, to provide more specific guidelines for athletic needs.

Discussion

Football is a demanding sport requiring repeated high-intensity interval performance, including speed endurance. With elevated carbohydrate oxidation rates, rapid glycogen depletion could negatively influence physical and mental performance. Furthermore, poor hydration status and/or high individual sweat rates might also contribute to player fatigue (1). In addition, the limitations imposed by player specific training demands, travelling, food timing strategies, and individual needs pre/post match could all contribute to getting nutritional practices correct.

It has been documented that good nutrition practices are important for energy metabolism, and performance efficiency (2). In terms of daily nutrition practices, total calories consumed as well as total carbohydrate and protein intakes are considered key to high-level performance. However, Total Energy Intake (TEI) for football athletes have been reported to be below recommended levels, with estimated TEI needs ranging from 3819 – 5185 kcal/day (3). This implies the need for nutritional assessment and player support.

Observations from Dietary Intake Assessment.
1)   Energy and Carbohydrate Intake:
A 4-day dietary recall was obtained on 16 players, with consideration given to total food/fluid intake over a 24-hour period, including portion size and meal timings. Average TEI was surprisingly low at 1998.6±412.2 kcal/d. It was noted that there was wide variance between players, with average TEI probably influenced by low food consumption on recovery days, and tendency to skip meals pre and/or post training. Such practices have been linked with injury prevalence (4), overreaching (5) and compromised power to weight ratios (6). Energy intake was therefore highlighted as a key area to address through additional meal planning and pre training loading.

Average daily carbohydrate intake for the 16 players interviewed was 228.0±66.5 g/d. It has been recommended that for team sports, daily carbohydrate intake guidelines should fall between 5-7g/kg/d (7). For an average 75kg player, this would require a daily carbohydrate intake of 375-525g/d. There was again diversity between players in terms of total carbohydrate intake; however, on average players were consuming 3g/kg/d. This is considerably less than demonstrated elsewhere (8). Increasing carbohydrate intake closer to 8g/kg/d has been shown to increase pre training muscle glycogen content, and extend the total time performing high-intensity work (9).

2)   *Protein and Fat Intake:*

The Dietary Reference Value (DRV) for protein is 55.5g/d (10). This would equate to approximately 0.74g/kg/d for a typical 75kg athlete. However, recommended daily protein intake for athletes, with reference to football has been stated as 1.4-1.7g/kg/d (11). Within this squad, protein intake was 130.4±49.6 g/d (equating to approximately 1.74g/kg/d for an average player weighing 75kg). Whilst this appeared to be sufficient, the variance between players indicated the need for individualized support. Indeed, some players consumed close to or below DRVs; which has been shown to accentuate exercise-induced immune suppression (12).

Average daily fat intake was 66.0±28.8 g/d, with 23.7±10.5 g/d coming from saturated fat (36% of total fat intake) and 10.5±4.4g/d coming from polyunsaturated fats (16% of total fat intake). The total fat intake was found to be approximately 30% of TEI, which falls within current population guidelines. It is questionable whether this amount is suitable for such athlete's warrants consideration, especially in light of both low energy and carbohydrate intakes.

3)   *Micronutrient Intake:*

With regards to micronutrients, as metabolic rate is increased, nutrient turnover is accelerated. This could lead to marginal states of acute nutrient deficiency (8). Magnesium, for example, is intrinsically involved in energy regulation, acting as a cofactor for various enzymes. Increased sweat rates lead to increased loses of magnesium and other electrolytes, potentially leading to increased functional demand. Other micronutrients have important functions as antioxidant cofactors and hence support post exercise free radical defence (13).

A summary of average daily micronutrient intake is shown in Table 1 in comparison to Recommended Nutritional Intakes (RNIs). The results indicated that players were on average consuming sufficient amounts of micronutrients in relation to population based RNIs. Whether these are appropriate for professional athletes warrants consideration in relation to suggested 'performance intakes', 'upper tolerance levels' and results from functional assessment. The degree of variance between players in terms of dietary intake should also be noted.

Observations from the Diagnostic Tests:

The NutrEval profile provides an overview of nutritional demands at a functional level. Assessment of amino acids, organic acids, essential fatty acids, oxidative stress, toxic elements and nutrient ratios from plasma and urinary sampling provide a potential means to overview gastrointestinal health, cellular energy production and vitamin/mineral demands (14,15).

Although average daily carbohydrate intake was reported to be below recommended guidelines, the habitual intake of refined sugars was relatively high. Interestingly, assessment of gastrointestinal dysbiosis was evident in 44.4% of the squad. This could relate to the relatively high glycemic index of assessed diets, and supported the recommendation to modify the glycemic load of player diets. There was also evidence of protein maldigestion, with 81.5% of the squad demonstrating elevated functional indices. Whilst this could relate to negative eating patterns (for example, inadequate chewing time, or rushed eating), protein maldigestion is indicative of digestive enzyme inefficiency, increased mucosal permeability and/or gastrointestinal dysbiosis.

Despite the fact the dietary analysis indicated normal RNIs for all selected antioxidant nutrients, 96.3% of the whole squad (and 88% of the 16 players assessed for dietary intake) were found to have moderate to high functional indices for antioxidant status. This potentially suggested that the physical demands for these athletes exceeded population guidelines. Additionally, 85.2% of the whole squad were found to have a moderate to high oxidative stress index (82% for the players assessed separately). Such information should be treated with consideration, but does indicate the potential need for increased wholefood antioxidant nutrients (15).

Plasma 25-hydroxycholecalciferol is accepted as a valid marker of vitamin D status (16). Diagnostic data revealed that 78.6% of the whole squad had below average levels of vitamin D (40.1±13.5ng/ml); and was similar when adjusted for those players undertaking dietary assessment (74.0%; with average levels of vitamin D being 38.5±12.9ng/ml). Compromised vitamin D status may negatively influence innate immunity, inflammatory cytokine cascades, and lead to suboptimal athletic performance (17). With average daily intake of vitamin D recorded at 3.8±3.6μg/d, there is indication that dietary vitamin D intake should be higher in athletes (18).

Functional indices for magnesium demand were elevated in 86% of the players assessed (and 80.8% for the whole squad), despite normal red blood cell (RBC) magnesium levels for all athletes. RBC magnesium has been shown to be a sensitive marker of magnesium deficiency (15); however, as nutrient turnover is likely to be increased with repetitive exercise, the functional test data may provide useful information for increased nutrient demands for such athletes.

Table 1: Selected micronutrient intakes (n=16) in relation to Recommended Nutritional Intake and NutrEval Indices.

NUTRIENT	Dietary Assessment Evaluation	Recommended Nutrient Intakes (RNIs)	NutrEval Indices (% of whole squad showing elevated nutrient demand indicated)
Vitamin A (μg/d) (Retinol equivalents)	691.9 ± 322.4	700	↑ antioxidant demand (96.3%)
Vitamin C (mg/d)	109.8 ± 81.0	40	
Vitamin E (mg/d)	5.4 ± 2.4	>4	
Vitamin D (μg/d)	3.8 ± 3.6	0-10	↓ plasma status* (78.6%)
Calcium (mg/d)	753.3 ± 241.9	700	Not assessed
Magnesium (mg/d)	311.0 ± 96.6	300	↑demand (80.8%)
Zinc (mg/d)	10.4 ± 3.0	9.5	Low demand (23.1%)
Iron (mg/d)	13.6 ± 6.0	8.7	↑demand (42.3%)
Copper (mg/d)	1.4 ± 0.7	1.2	↑demand (38.5%)
Selenium (μg/d)	90.1 ± 49.6	75	↑ antioxidant demand (96.3%)
Vitamin B$_6$ (mg/d)	3.02 ± 1.00	1.4	↑demand (63.0%)
Vitamin B$_{12}$ (μg/d)	5.8 ± 3.3	1.5	Normal (<10%)
Folate (μg/d)	252.9 ± 85.9	200	Low demand (25.9%)

*Plasma 25-hydroxycholecalciferol

Conclusions

In preparation for the 2010 South Africa World Cup, the use of combined diagnostic and dietary assessment provided useful information on player/squad nutritional needs. Whilst there are limitations imposed with the use of such methods, and care should be taken when interpreting data, the results indicated the need to address factors such as total energy and carbohydrate intake, antioxidant nutrients, and vitamin/mineral demands.

Observations from both assessments highlighted the importance of regular monitoring of individual and team nutritional intake, as well as sourcing both robust and practical evaluation methods. Awareness of essential dietary practices such as frequent meal patterns, portion size, protein-carbohydrate ratios, food timing, and appropriate recovery strategies were addressed as part of the follow up programme as a means to support athletic performance.

References

1. Coyle EF. Fluid and fuel intake during exercise. *J. Sports Sci.* 2004;**22**:39-55.
2. Kirkendall DT. Effects of nutrition on performance in soccer. Med. Sci.   Sports Ex. 1993;**25**:1370–1374.
3. Leblanc JC, LeGall F, Grandjean V, Verger P. Nutritional intake of French soccer players at the Clairefontaine Training Center. *Int. J. Sport Nutr. Exer. Metab.* 2002;**12**:268–280.
4. Eichner ER. Overtraining: Consequences and prevention. *J. Sports Sci.* 1995;**13**:41–48.
5. Achten J, Halson SL, Moseley L, Rayson MP, Casey A, Jeukendrup AE. Higher dietary carbohydrate content during intensified running training results in better maintenance of performance and mood state. *J. Appl. Physiol.* 2004;**96**:1331–1340.
6. Maughan R. Role of micronutrients in sport and physical activity. *Br. Med. Bulletin.* 1999;**55**(3):683-690.
7. Burke LM, Loucks AB, Broad N. Energy and carbohydrate for training and recovery. *J. Sports Sci.* 2006;**24**(7):675-685.
8. Reeves S, Collins K. (The nutritional and anthropometric status of Gaelic football players. *Int. J. Sport Nutr. Exer. Metab.* 2003;**13**:539-548.
9. Balsom PD, Wood K, Olsson P, Ekblom B. Carbohydrate intake and multiple sprint sports: with special reference to football (soccer). *Int. J. Sports Med.* 1999;**20**:48-52.
10. Geissler C, Powers H. *Human Nutrition* (11th Ed.) Edinburgh: Elsevier.
11. Lemon PWVR. Protein requirements of soccer. J Sports Sci. 2005;**12**:S17-22.
12. Gleeson M, Nieman DC, Pedersen BK. Exercise, nutrition and immune function. *J. Sports Sci.:* 2004;**22**:115-125.
13. Sen, CK. Antioxidants in Exercise Nutrition. *Sports Med.*; 2001;**31**(13):891-908.
14. Genova Diagnostics *NutrEval Interpretation Guide.* London: Genova Diagnostics Ltd; 2008.
15. Lord RS, Bralley JA. *Laboratory Evaluations for Integrative and Functional Medicine* (2nd Ed.). Duluth, Georgia: Metametrix Institute; 2008.
16. Jamison J. *Clinical Guide to Nutrition and Dietary Supplements in Disease Management.* Australia: Churchill Livingstone; 2003.
17. Willis KS, Peterson NJ, Larson-Meyer DE. Should We Be Concerned About the Vitamin D Status of Athletes? *Int. J. Sport Nutr. Exer. Metab.* 2008;**18**:204-224.
18. Vieth R. Vitamin D supplementation, 25-hydroxyvitamin D concentrations, and safety. *Am. J. Clin. Nutr.* 1999;**69**:842–56.

Commentary: Chris Barnes, Australian Institute of Sport.

Justin Roberts presents an interesting report on the nutritional intake of elite soccer players in the run up to the 2010 World Cup, coupled with an appraisal of associated Glycaemic Index and cellular function. The reported sub-optimal calorie intake concurs with the findings of many other studies of a similar nature, both in the UK and abroad. Whilst the use of dietary recall diaries may have its limitations, the consistency of findings raises cause for concern for team sports athletes preparing for competitions such as the World Cup.

It is interesting to note that the shortfall in calorie intake would appear to be in carbohydrates, and more specifically complex carbohydrates for many players, with other macronutrient intakes being within what would be considered adequate for their population. It is a shame that there is no reference to the time of the season, or phase of training cycle that the food diaries represent, which would have provided a little more context to the case study.

When prescribing nutrient intake strategies for soccer players, consideration should be given to the potential huge variation in their weekly training / match structure. Depending on circumstances, players could participate in one, two or, in exceptional circumstances, three matches within a 7-day period.

Given that a 90-minute game of soccer can result in muscle glycogen levels dropping by 40% to 90% (1), then restoration of glycogen stores through appropriate nutrient intake is of paramount importance.

Additionally, daily training volume and intensity can vary markedly in accordance with the phase of the season and aims of sessions. For example, volume, represented by total and high-intensity locomotive distances covered may range between 30% and 80% of individual match performances. Similarly, intensity, represented by total and high-intensity distances covered per unit time may range between 30% and 120% of individual match performances. Such variation in training / match demands has clear implications on both a qualitative and quantitative level for nutrient prescription. Add to this the different positional demands of the sport with, for example, wide midfield players covering double the distances at high-intensity than centre halves (2) then the job of the sports nutritionist in providing a prescription which is appropriate to individual, positional and circumstantial needs becomes even more challenging.

Individualised and periodic nutrient intake strategies for elite team sports athletes are essential to optimise both performance and recovery, and to protect hormonal, metabolic and immune function (3). Regular monitoring of intake and associated functional indices of health are an important component of any support programme for elite team sports athletes.

References

1. Bangsbo J, Iaia FM, Krustrup P. Metabolic response and fatigue in soccer. International Journal of Sports Physiology and Performance. 2007:**2**(2):111-27.
2. Bradley PS, Sheldon W, Wooster B, Olsen P, Boanas P, Krustrup P. High-intensity running in English FA Premier League soccer matches. Journal of Sports Sciences. 2009;**27**(2):159-68.
3. Burke LM, Loucks AB, Broad N. Energy and carbohydrate for training and recovery. Journal of Sports Sciences. 2006;**24**(7):675-685.

CHAPTER 19

Coach-Led Exercise Training Programs Aimed at Preventing Lower Limb Injuries in Players: Should the Focus be on Injury Prevention Gains, Likely Performance Benefits or Both?

Caroline Finch
Australian Centre for Research into Injury in Sport and its Prevention (ACRISP), Federation University Australia, Ballarat, Victoria, Australia.

Acknowledgements

The extracted data are from the Preventing Australian Football Injuries through eXercise (PAFIX) study which was funded by a National Health and Medical Research Council (NHMRC) project grant (ID: 400937). An NHMRC Principal Research Fellowship funded Professor Caroline Finch (ID: 565900). ACRISP is one of the International Research Centres for Prevention of Injury and Protection of Athlete Health supported by the International Olympic Committee (IOC).

Vignette

With the accumulating international evidence from both randomised controlled trials and biomechanical studies that indicate lower limb injuries (LLI) in sport can be prevented through targeted exercise training (1-6), it was decided to trial the effectiveness of an exercise training program specifically for Australian Football players. There was considerable evidence about what the content of the exercise training program should involve. A program was developed that incorporated a structured warm-up, balance training, side-stepping/cutting skills and jump/landing training components (7). In addition to the content of a training session, we also investigated issues related to the delivery. In contrast, it was not known whether players would actually do the exercises if their coaches delivered this new evidence-based training program to them during standard training sessions. It was important to determine this before the training program was implemented because the only interventions that will prevent injuries are those which are adopted by players (8).

Australian Football players from nine senior clubs/teams were surveyed about their knowledge about LLI-prevention and their opinions about the value of specific training program components to prevent LLIs (9). Because we considered it important that players might be more motivated to undertake specific exercise training if there was a clear performance benefit, rather than only an injury prevention benefit, they were also asked about that.

Whilst three-quarters of the 374 surveyed players agreed that doing specific exercises during training would reduce their risk of LLI and would be willing to undertake them, 64% also thought that training should focus more on improving game performance than injury prevention (9). Players were most supportive of improving kicking and ball handling skills for performance and undertaking warm up runs and cool-downs for injury prevention. Fewer than three-quarters of all players believed that balance, landing or cutting/stepping training had LLI prevention benefits.

Although it is the players themselves who need to participate in exercise training for LLI prevention, most sessions are designed to be delivered by coaches during the standard training sessions. Coaches, therefore, have a significant role to play in the uptake of injury prevention measures by their players (10). For this reason, the head coaches from the same nine clubs/teams were also surveyed and asked similar questions.

All coaches wanted to prevent LLI in their players and were most receptive towards implementing specific training program components if they concurrently improved player performance and maximised injury prevention (11). However, it was also clear that most coaches were not aware of the latest scientific findings in relation to exercise training programs for LLI prevention, including what specific training components to emphasise.

The discussion below summarises both the views of players and coaches about the relative benefits of performance gains and injury prevention that are associated with different training program components.

It also discusses the implications for the future delivery of training program components for LLI prevention.

Discussion

There is a growing recognition for well-designed studies that examine the effectiveness of interventions conducted within the real-world context of sports delivery. However, despite the importance of such studies, few studies have been conducted and published (12). It has only been recently that direct empirical evidence of the detrimental impact that non-adoption of specific exercise training program components has on intervention effectiveness has become available (5, 6). It is now clear that broad LLI prevention across many sports has been limited in its success and reach because little prior attention has been given towards understanding implementation issues, including barriers and facilitators to sustainable programs (12).

Before training that have been shown to be efficacious in terms of injury prevention and improving performance, a detailed understanding of the context of sport delivery is needed (10). For example: "too what extent do coaches currently give adequate attention to the development of training skills that are most likely to reduce the risk of lower limb injuries in their players?" and "What are players' prevailing attitudes and beliefs in relation to lower limb injury causes, predisposing factors and preventive measures?" Answers to these questions are needed for both performance and health reasons. In team sports settings where the coach leads training sessions for a group of players, information about both coach and player views about the likely injury prevention and performance value of various training program components, would be useful in the development of such sessions.

Most of the surveyed players understood that specific training exercise programs have a role in LLI prevention (9). Figure 1 shows comparative opinions about the value of various training program components for performance gains or injury prevention. There was strong agreement that most training session components should focus on improving performance; most players agreed that undertaking specific exercises would prevent LLI but, apart from warm-up, they generally thought the benefits were greater for performance. Based on this result, it is suggested that there is a need to educate players about the LLI prevention benefits of some important training session components such as skills training, ball handling and jump/landing technique, as many only rated them as highly important for performance. Surveyed players also indicated that they did not want injury prevention to be the major focus of their training sessions, an opinion consistent with their higher rankings of performance benefits.

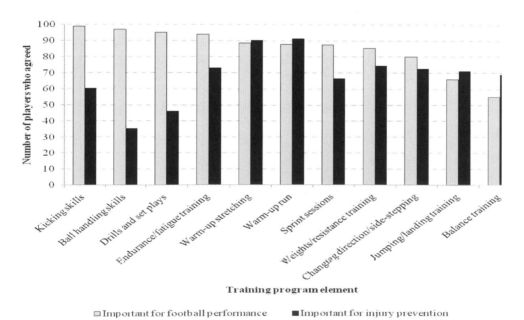

Figure 1: Players' views about the relative importance of various training program components for different purposes (n=374). (Data extracted from (9)).

A major implication of these findings that if there are no clear concurrent performance outcomes then LLI prevention programs should not be promoted as being performance enhancing as players will not be supportive of them. We have therefore recommended that injury prevention exercises delivered as components of broader training programs are acknowledged as such, and that they should have a shorter duration than training components aimed at performance and game development (9). Moreover, as players have strongest LLI prevention support for the warming-up type components, it may be advisable to place activities that are purely aimed as injury prevention towards the start of training sessions.

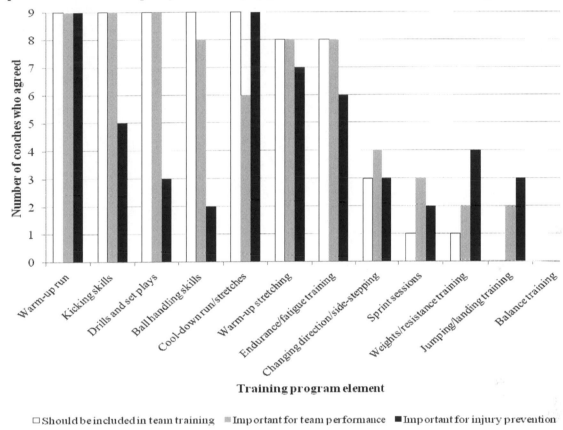

Figure 2. Coaches' views about the relative importance of various training program components for different purposes (n=9 coaches surveyed). (Data extracted from (11))

Figure 2 shows the views about the same training components as expressed by the coaches of the surveyed players. Overall, and perhaps not surprisingly, these Australian Football coaches ranked injury prevention lower in priority than both general training session needs and team performance goals. They highly rated the importance of the warm-up and cool-down for both injury prevention and performance. However, they seem to be unaware of latest evidence about the benefits of balance training, and improving landing/side-stepping/change-of-direction skills (1, 2, 13-15).

The implications of this are that coaches will design their training programs to favour those components they are most familiar with and not include other components that the scientific evidence now makes clear are of benefit for LLI prevention. There is a clear is a need for improved translation of the latest scientific evidence about the most effective injury prevention strategies so that it directly informs coaches' knowledge base and practice (11).

Conclusions

Accumulated scientific knowledge now exists about how best to prevent lower limb injury, including the key role that specific and targeted training programs can play in this. However, other research shows that there is poor uptake of preventive measures and low adherence or compliance to recommended exercise training programs, suggesting that more effort needs to be given towards identifying how best to deliver such injury prevention programs to athletes. Information about how to best promote, market or deliver injury prevention programs to ensure their wide-scale adoption and implementation is now needed, as it is only those safety recommendations which are adopted as standard sporting practice that will actually prevent injuries (8, 12). Importantly, this information is a necessary precursor to the development of an effective injury prevention program, to be delivered during coach-led training sessions (10).

Neither players nor coaches seem to be aware of the latest evidence about the role of specific training components, specifically landing/cutting and balance training, for LLI prevention. There is now increased onus on sports injury prevention researchers to translate their scientific evidence directly to the sports settings in which it needs to be applied, to ensure that coaches are well-informed and players are best protected.

There is a general consensus that team training sessions are an ideal setting for coaches to pass on injury prevention information to their players. Moreover, specific game performance elements can be practised without full game-level stressors, including increased injury risks. The optimal preventive approach would be to deliver targeted exercise training components as part of routine football training program activities. This would integrate them as standard practice, as a means of associating them with the broader and currently more accepted training benefits.

References

1. Verhagen E, van der Beek A, Twisk J, Bouter L, Bahr R, van Mechelen W. The effect of a proprioceptive balance board training program for the prevention of ankle sprains: a prospective controlled trial. *Am J Sports Med.* 2004;**32**:1385-1393.
2. Olsen O, Myklebust G, Engebretsen L, Holme I, Bahr R. Exercises to prevent lower limb injuries in youth sports: cluster randomised controlled trial. *Br Med J.* 2005;**330**:449.
3. Verrall G, Slavotinek J, Barnes P. The effect of sports specific training on reducing the incidence of hamstring injuries in professional Australian Rules football players. *Br J Sports Med.* 2005;**39**:363-368.
4. Gabbe B, Branson R, Bennell K. A pilot randomised controlled trial of eccentric exercise to prevent hamstring injuries in community-level Australian football. *J Sci Med Sp.* 2006;**9**:103-109.
5. Engebretsen AH, Myklebust G, Holme I, Engebretsen L, Bahr R. Prevention of injuries among male soccer players: a prospective, randomized intervention study targeting players with previous injuries or reduced function. *Am J Sports Med.* 2008;**36**:1052-1060.
6. Steffen K, Myklebust G, Olsen O, Holme I, Bahr R. Preventing injuries in female youth football – a cluster-randomized controlled trial. *Scand Journal Med Sci Sp.* 2008;**18**:605-614.
7. Finch C, Lloyd D, Elliott B. The Preventing Australian Football Injuries with eXercise (PAFIX) study - a group randomised controlled trial. *Inj Prev.* 2009;**15**(e1. Available from 10.1136/ip.2008.021279).
8. Finch C. A new framework for research leading to sports injury prevention. *J Sci Med Sp.* 2006;**9**:3-9.
9. Finch C, White P, Twomey D, Ullah S. Implementing an exercise training program to prevent lower limb injuries – considerations for the development of a randomised controlled trial intervention delivery plan. *Brit J Sports Med.* 2011;**45**:791-795.
10. Finch C, Donaldson A. A sports setting matrix for understanding the implementation context for community sport. *Brit J Sports Med.* 2010;**44**:973-978.
11. Twomey D, Finch C, Roediger E, Lloyd DG. Preventing lower limb injuries: is the latest evidence being translated into the football field? *J Sci Med Sp.* 2009;**12**:452-456.
12. Finch C. No longer lost in translation – the art and science of sports injury prevention implementation research *Br J Sports Med.* 2011;BJSM Online First, published on June 22, 2011 as 10.1136/bjsports-2011-090230.
13. Cochrane J, Lloyd D, Besier T, Elliott B, Doyle T, Ackland T. Training affects knee kinematics and kinetics in cutting maneuvers in sport. *Med Sci Sports Exerc.* 2010;**42**:1535-1544.
14. Dempsey AR, Lloyd DG, Elliott BC, Steele JR, Munro BJ. Changing sidestep cutting technique reduces knee valgus loading. *Am J Sports Med.* 2009;**37**:2194-2200.
15. Lloyd D. Rationale for training programs to reduce anterior cruciate ligament injuries in Australian football. *J Ortho Sp Phy Ther.* 2001;**31**:645-654.

Commentary: Justin Roberts, University of Hertfordshire, UK

This case study concisely articulated by Professor Finch outlines a number of current and key issues facing contemporary sports science and health professionals working with high-level performance. The first is the central need to undertake appropriate and up to date epidemiological field research to ascertain more precise information pertaining to specific training and performance injury occurrence. With this, it is evident that clarity of dissemination of such information should be apparent at all tiers of the sports science, coaching and medical support structure particularly within team sports.

Furthermore, the need to undertake more "specific" and evidence-based intervention programmes for high level performance is congruently identified by Professor Finch as a means to enhance training effectiveness and minimise injury occurrence. The concept of "targeted exercise training" reflects one of the key principles of performance training, that of training specificity. However, within this case study, the evidence infers the need utilise training specificity and develop training session that include drills that progressively reinforce, and/or support, performance demands whilst enhancing neuro-muscular, physiological and metabolic adaptations designed to minimise or even prevent the likelihood of LLI occurrence.

A second key issue identified within the case study, is the underlying requirement for athlete engagement and adoption of such targeted exercise training. Within this, there is the implication of player knowledge/awareness leading to both understanding of 'periodization flexibility' and the correlation of training loading to progressive performance gains. The dissemination of research evidence during pre and/or off-season training camps to all levels (coaching staff, medical staff, players) may therefore be a critical component of player adoption and correct implementation of targeted exercise training.

The data presented by Professor Finch indicates an apparent disconnect between training for acute performance gains and training to support longer term adaptations leading to higher level performance. This is a similar observation with nutritional interventions, in that often world class performance gains are apparent when a structured, monitored and regulated programme is implemented in a specific, periodised manner over a typical season. However, the information presented from player surveys does highlight the importance of blending components of both conditioning and LLI prevention training during periods of standard training sessions when players are likely to be more responsive. Additionally, it is evident there is a need to transparently blend LLI prevention training alongside performance specific training for player adherence and benefit.

Whilst the sample size was relatively small, it was interesting to note that areas such as side-stepping, sprinting, resistance exercises and balance skills were poorly rated amongst the coaches as being important for performance, as well as injury prevention. Potentially, this observation may be different during various phases of the competitive season. Additionally, the need to enhance player motivation and 'readiness to train' leading to improved overall player availability for regular matches may be a significant influential factor from the team managers perspective. This could be a primary factor that may impact on both player and coach's interaction with, and inclusion of, scientific principles. Overall however, the case study succinctly demonstrates the importance of the sports science/health professionals' ability to disseminate clear, concise, evidence-based and current research information that can be effectively applied by coaches to enhance player performance whilst minimising LLI risk.

Part III: The Proactive Model: Providing on-going support and developing interdisciplinary teams

CHAPTER 20

Making the Weight: Case-studies from Professional boxing

James P Morton and Graeme L Close
Research Institute for Sport and Exercise, Liverpool John Moores University, Liverpool, UK.

Introduction: background to issue

Professional boxing is categorised into weight classes intended to promote fair competition by matching opponents of equal stature and body mass (commonly referred to as 'weight' within the sport). Boxing is a sport that has its own tradition and culture in relation to weight making practices. Many boxers achieve their target weight via the combination of acute and chronic means that involve severe energy restriction and dehydration (1). The latter weight-making method is common in the days preceding the weigh-in and is known as 'drying out'. Depending on the stage of the boxer's career, the number of contests per annum usually ranges from one to six. Typical durations of training camps for each contest ranges from 6-12 weeks and it is not uncommon for boxers to commence training with considerable weight to lose (see Table 1).

Table 1. Body composition analysis as assessed via dual-energy x-ray absorptiometry (DXA; QDR Series Discovery A, Holognic Inc, Beford, MA) for three professional boxers at the onset of their training camps. In these cases, the athletes were only competing on average three times per year and thus the inactivity in-between training camps resulted in the boxers commencing training with considerable weight loss requirements. Note that for Boxer 1 and 3, the required weight loss could only be achieved via a combination of both fat and lean mass loss.

Boxer	Weight Division	Body Mass (kg)	Body Fat (%)	Fat Mass (kg)	Lean Mass (kg)	Required Weight Loss (kg)	Duration to Make Weight (weeks)
1	Super-featherweight (59 kg)	68.3	12.1	8.5	56.6	9.3	12
2	Light-heavyweight (81 kg)	92.0	18.1	17.2	73.1	11.0	9
3	Flyweight (51 kg)	60.1	17.7	10.9	47.9	9.1	8

Although data concerning weight-making practices of professional boxers is scarce (2), research examining amateur boxers reported weight losses of 3-4 kg in the week preceding competition (1). The acute dehydration that is common to such weight losses impair performance as evidenced by reduced punching force (3), whilst dehydration and energy restriction carry obvious health risks. Indeed, reductions in energy and fluid intake during training (and in the weeks and days prior to competition) may increase infection risk (4), induce fluctuations in mood state (5) and compromise training intensity (6). Furthermore, increased cardiovascular and thermoregulatory strain could induce injury and, in extreme cases cause fatality (7).

This chapter provides an overview of ongoing support work (over a 5 year period) adopting a more structured and scientific approach to making weight for a 25 year old professional male boxer competing in the 57 kg featherweight division (Boxer 1 from Table 1). We begin by presenting an overview of the athlete's sporting background followed by supporting scientific rationale and outcomes of our chosen nutritional interventions. Data from the initial support period has been published previously (2) and we refer to these data (where appropriate) so as to provide a full case history of our ongoing support work.

Presentation of athlete and overview of sporting history

Boxer 1's achievements include junior and senior national amateur champion titles as well as a junior Olympic gold medal. He turned professional at age 18 and soon achieved success acquiring a national featherweight title and later held a version of the world featherweight title. The athlete's usual approach to making weight relied on a 6-8 week training camp comprising a daily diet consisting of one meal (usually consumed at lunch time and consisting of one large sandwich and a diet cola), daily use of sweat suits during training, no consumption of food or drink for one to two days preceding the weigh-in (he also used sweat suits and low-intensity exercise in the hour preceding the weigh-in) and re-fuelling strategies between weigh-in and competition consisting of high fat based foods such as confectionary products and fried foods e.g. bacon, eggs, sausages etc. The athlete revealed he was now struggling to make the featherweight limit and had decided to move up in weight to the 59 kg super-featherweight division.

Overview of initial nutritional and conditioning intervention

Based on assessment of body composition (see Table 1) and despite moving up in weight, we initially realised the 59 kg weight limit could only be achieved via a combination of fat and lean mass loss (see (2)) for a more detailed overview of initial athlete assessment). To achieve this target weight loss of 0.5 -1 kg over a 12-week period, we opted to attain an energy deficit via a combination of restricted energy intake (restricted to approximate values of resting metabolic rate, RMR) and increased energy expenditure. Relative to the athlete's habitual energy intake, our initial nutritional intervention consisted of reduced fat and carbohydrate (CHO) intake concomitant with increased protein intake (see (2) for example, meal plans where the athlete adhered to a daily diet approximately equivalent to his resting metabolic rate: 6-7 MJ. 40% CHO, 38% Protein, 22% Fat). The rationale for reduced CHO intake (2-3 g/kg body mass) was to enhance lipid oxidation and stimulate gluconeogenesis to achieve lean mass loss (Howarth et al. 2010)(8). In order to not induce the latter process at too great a rate and to maintain daily calorie intake at our intended level, we also increased protein intake to 2-2.5 g/kg body mass which included the provision of protein ($\approx$ 20 g) in close proximity to the onset and completion of exercise so as reduce protein degradation and promote protein synthesis (9,10,11). Fat intake was restricted to < 1 g/kg and was largely achieved via consumption of unsaturated fats from fish sources and oils. CHO foods were consumed from low to moderate glycemic index (GI) based foods to minimise the suppression of lipolysis in the post-prandial period and during subsequent training sessions (12). The athlete also consumed a multivitamin supplement providing 100% of the RDA to potentially maintain immune function in conditions when energy intake is compromised . Fluid intake was restricted to water or low calorie flavoured water and was consumed *ad libitum* throughout each day though the athlete was instructed to consume the necessary fluid to re-hydrate immediately after each training session (13). The athlete's training programme comprised fasted runs at moderate-intensity (6 am) so as to maximise lipid oxidation (14,15) and promote oxidative training adaptations (16), boxing-specific sessions (11 am) and finally, strength and conditioning sessions (5 pm) three times per week. The scheduling of resistance training at this time was to promote circadian influences on strength as well as to avoid any molecular interference effect associated with concurrent training in close proximity (17).

Outcome of the initial intervention

In our initial support period (2), average body mass loss was 0.9 ± 0.4 kg per week equating to a total loss of 9.4 kg (a weight loss representing a decrease in percent of body fat from 12.1 to 7.0%). Following the weigh-in, the athlete consumed a high CHO diet (12 g/kg body mass) supported by appropriate hydration strategies and entered the ring 30 hours later at a weight of 63.2 kg. Our initial intervention represented a major change to the athlete's habitual weight making practices and did not rely upon any form of intended dehydration during the training period or preceding weighing-in. Consistent with other authors (18), we also observed that the provision of a high protein diet and structured resistance training was relatively successful at maintaining lean mass (despite being in a daily energy deficit) and it was only until considerably low body fat levels were reached did lean mass show the necessary decline to make the 59 kg weight limit (see Figure 1). He also reported that this was the easiest he had ever made the weight despite energy and fluid intake four to six times per day.

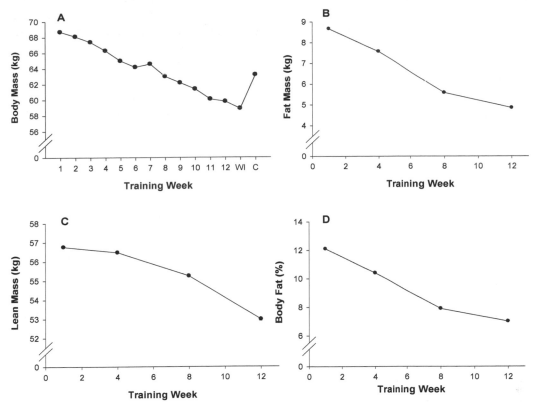

Figure 1. Changes in Boxer 1's (A) body mass, (B) fat mass, (C) lean mass and (D) per cent of body fat throughout the 12 week intervention period to make weight for the 59 kg super-featherweight division. WI = weigh-in, C = competition (data adapted from Morton et al. 2010)(2).

Ongoing support work

Despite the new approach to making weight and also one further contest at the 59 kg limit (both of which were defeats), the athlete retired from the sport citing boredom with training, competition and the repeated requirement to make weight as contributing factors (in effect, the athlete reported he had fallen out of love with the sport). However, after a 9 month retirement period in which he was involved in coaching young novice boxers, he re-developed a passion for the sport and decided to box competitively again. At this time, he also changed coaches and his new coaching team agreed he should compete at the lightweight limit of 61.4 kg, almost 4.5 kg heavier than the featherweight division that he had competed in most of his career. At this new weight division, the athlete was thus able to increase daily carbohydrate intake to 3-5 g/kg (depending on the duration and phase of the training camp) and maintain protein and fat intake at 2-2.5 g/kg and < 1g/kg, respectively. With this increased energy intake (and continued integration of weight training into his training plans) and an intentional acute dehydration of 1-1.5 kg prior to weigh-in, lean muscle mass is now typically 2-2.5 kg greater at the time of competition than when he boxed at super-featherweight. The athlete now reports that mentally, his training is focused more on improving boxing specific technique and fitness as opposed to becoming pre-occupied with making weight. He also feels that he is at the strongest and fittest point of his career. With the continual development of trust between the boxer and support staff, we now also regularly incorporate various supplement strategies (e.g. β-alanine, HMB, CLA, omega 3 fatty acids, BCAAs, vitamin D) in an attempt to promote training adaptation, recovery and performance. After appropriate re-fuelling and hydration following weigh-in, he now usually enters the ring at a fighting weight of 65-66 kg, a weight that he and his coach feels he performs best in training and sparring. Since his comeback, he has won numerous national and international titles and the athlete remains convinced that boxing at this heavier weight (which he now believes is his natural boxing weight), is one of the main contributing factors.

Conclusions

It is difficult to provide definitive recommendations for making weight in this chapter, owing to the fact that every boxer presents a different scenario in terms of RMR, target weight loss, daily training energy

123

expenditure, time to achieve target weight etc. However, we have recently published guidelines for combat athletes (19) encouraging a strategy focusing on increased protein intake in combination with resistance training (so as to maintain lean mass in the face of daily energy deficits), reduced CHO availability (and the reliance upon LGI CHO sources to promote lipid oxidation), as well as reduced intake of those foods that are simultaneously high in sugar and saturated fat intake. Additionally, emphasis should be placed upon coach and athlete education so as to develop a training culture, which promotes adequate hydration before, during and after training (see Langan-Evans et al. 2011 for a detailed discussion of these guidelines).

In addition to changes in *composition* and *quantity* of energy intake, it is also important that *timing* of energy intake is aligned to the structure of the daily training schedule in order to promote lipid oxidation, training adaptation and recovery (see Table 2). We have used similar principles with professional boxers ranging from flyweight to heavyweight (in the latter case, to change body composition and not necessarily mass) and observed positive results (see Figure 2). The success of these interventions are underpinned through continual education of coach and athlete and the willingness of both parties to adopt novel practices despite being unfamiliar to boxing culture. Given the lack of research in this area, we consider it vital that similar case-study type accounts from other weight-making sports are published in the scientific literature (e.g. see reference 20). Only through sharing such information can the safety and performance of such athletes be enhanced.

Table 2. Overview of guidelines for timing and composition of nutritional and fluid intake in relation to the structure of the daily training schedule. In this case, strategies are included for a boxer performing 3 training sessions per day (this usually represents a boxer's most intense training day and such days typically only occur 2-3 times per week). Note that quantities of foods are not disclosed owing to the need for formulating individualised interventions.

Time	Training Session and/or Nutritional & Fluid Intake	Training and /or Nutritional Aims
630-715 am	Moderate-intensity steady state run undertaken in fasted state accompanied with appropriate fluid intake	Maximise lipid oxidation and promote hydration
730 am	Moderate CHO / moderate protein / low fat breakfast with appropriate fluid intake	Promote *some* restoration of liver and muscle glycogen and protein synthesis as well as re-hydration
10 am	Low CHO / moderate protein and low fat snack	Promote CHO availability and protein synthesis
11 am – 12.30 pm	Sport-specific training session accompanied with appropriate fluid intake	Development of sport-specific fitness / technique and promote hydration
1 pm	Moderate CHO / moderate protein and low fat lunch accompanied with appropriate fluid intake	Promote *some* restoration of liver and muscle glycogen and protein synthesis as well as re-hydration
4 pm	Moderate protein intake	Stimulate protein synthesis prior to strength and conditioning session
4.30-5.30 pm	Strength and conditioning training session accompanied with appropriate fluid intake	Development of sport-specific aspects of strength and conditioning and promote hydration
530 pm	Moderate CHO / moderate protein and low fat snack (or recovery drink) accompanied with appropriate fluid intake	Promote *some* restoration of liver and muscle glycogen and protein synthesis as well as re-hydration
7 pm	Low CHO / moderate protein and low fat dinner	Promote protein synthesis and hydration as well as minimising evening fat storage
10 pm	Moderate protein intake	Promote protein synthesis prior to sleeping

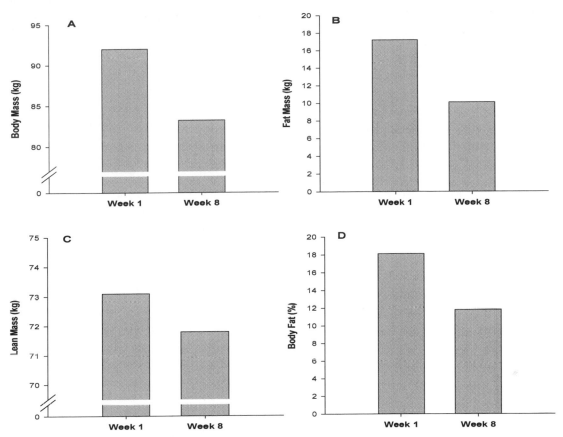

Figure 2. Changes in Boxer 2's (A) body mass, (B) fat mass, (C) lean mass and (D) per cent of body fat throughout a 9 week intervention period to make weight for the 81 kg light-heavyweight division. Data are presented for week 1 and 8 of the intervention period.

References

1. Hall CJ, Lane AM. Effects of rapid weight loss on mood and performance among amateur boxers. *Bri J of Sports Med.* 2001;**35**:390-395.
2. Morton JP, Sutton L, Robertson C, MacLaren DPM. Making the weight: a case-study from professional boxing. *Int J of Sports Nutn and Exe Met.* 2010;**20**:80-85.
3. Smith M, Dyson R, Hale T, Hamilton M, Kelly J, Wellington P. The effects of restricted energy and fluid intake on simulated amateur boxing performance. *Int J of Sports Nut and Exe Met.* 2001;**11**:238-247.
4. Gleeson M. Immune function in sport and exercise. *J of App Phys.* 2007;**103**:693-699.
5. Achten J, Halson SL, Moseley L, Rayson MP, Casey A, Jeukendrup, AE. Higher dietary carbohydrate content during intensified running training results in better maintenance of performance and mood state. *J of App Phys.* 2004;**96**:1331-1340.
6. Yeo WK, Paton CD, Garnham AP, Burke LM, Carey AL, Hawley JA. Skeletal muscle adaptation and performance responses to once a day versus twice every second day endurance training regimens. *J of App Phys.* 2008;**105**:1462-1470.
7. Centres for disease Control and Prevention. Hyperthermia and dehydration related deaths associated with intentional rapid weight loss in three collegiate wrestlers. *J of the Ame Med Assoc.* 1998;**279**:824-825.
8. Howarth KR, Phillips SM, MacDonald MJ, Richards D, Moreau NA, Gibala MJ. Effect of glycogen availability on human skeletal muscle protein turnover during exercise and recovery. *J of App Phys.* 2010;**109**:431-438.
9. Howarth KR, Moreau NA, Phillips SM, Gibala MJ. Coingestion of protein with carbohydrate during recovery from endurance exercise stimulates skeletal muscle protein synthesis in humans. *J of App Phys.* 2009;**106**:1394-1402.

10. Breen L, Philp A, Witard OC, Jackman SR, Selby A, Smith K, Baar K, Tipton KD. The influence of carbohydrate-protein co-ingestion following endurance exercise on myofibrillar and mitochondrial protein synthesis. *J of Phys*. 2011;**598**:4011-4025.

11. Coffey VG, Moore DR, Burd NA, Rerecich T, Stellingwerff T, Garnham AP, Phillips SM, Hawley JA. Nutrient provision increases signalling and protein synthesis in human skeletal muscle after repeated sprints. *Euro J of App Phys*. 2011;**111**:1473-1483.

12. Wee L-S, Williams C, Tsintzas K, Boobis L. Ingestion of a high glycemic index meal increases muscle glycogen storage at rest but augments its utilization during subsequent exercise. *J of App Phys*. 2005;**99**:707-714.

13. Shirreffs S, Taylor AJ, Leiper JB, Maughan RJ. Post-exercise rehydration in man: effects of volume consumed and drink sodium content. *Med and Sci in Sports and Exe*. 1996;**28**:1260-1271.

14. Horowitz JF, Mora-Rodriguez R, Byerley LO, Coyle EF. Lipolytic suppression following carbohydrate ingestion limits fat oxidation during exercise. *Ame J of Physio*. 1997;**273**:E768-E775.

15. Van Loon LJ, Greenhaff PL, Constantin-Teodosiu, D, Saris WH, Wagenmakers AJ. The effects of increasing exercise intensity on muscle fuel utilisation in humans. *J of Phys*. 2001;**536**:295–304.

16. Van Proeyen, K. Szlufcik, K. Nielens, H. Ramaekers, M & Hespel, P. (2011). Beneficial metabolic adaptations due to endurance exercise training in the fasted state. *J Appl Physiol, 110*, 236-245.

17. Coffey, V.G., Jemiolo, B., Edge, J., Garnham, A.P., Trappe, S.W. & Hawley, JA. Effect of consecutive repeated sprint and resistance exercise bouts on acute adaptive responses in human skeletal muscle. *Ame J of Physio*. 2009;**297**:R1441-1451.

18. Mettler S, Mitchell N, Tipton, KD. Increased protein intake reduces lean body mass loss during weight loss in athletes. *Med and Sci in Sports and Exe*. 2010;**42**:326-337.

19. Langan-Evans, C., Close, G.L. & Morton, J.P. Making weight in combat sports. *Stren and Cond J*. 2011; **33:** 25-39.

20. Wilson, G, Chester N, Eubank M, Crighton B, Drust B, Morton JP, Close GL. An alternative dietary strategy to make weight while improving mood, decreasing body fat and not dehydrating: a case-study of a professional jockey. *Int J of Sports Nut and Exe Met*. 2010;**22**:225-231.

Commentary: *Andy Lane*, University of Wolverhampton, UK.

The case study outlines a successful weight-making intervention with a professional boxer. In reviewing the case study, the first issue that stands out is the athlete's approach to weight making before the intervention. He ate one meal per day of dubious nutritional quality whilst engaging in heavy training. He would fast for two days before the weigh-in. The degree of self-control to manage this diet is impressive. He would experience intense hunger and most likely feel tired and lacking energy at a time when the opposite state was required. Following the weigh-in, all barriers to self-control are let down with the hope that he could feel strong and energetic before the contest.

The intriguing part of work such as this is the negotiation process. The boxer and his coach will hold mixed views on their existing strategy. They will hold a positive belief that it worked previously and appears successful for others whilst acknowledging, by the fact that external help has been requested, that change is needed. The boxer and his support team would hold beliefs on how severe this change might need to be, and the approach of identifying lean body mass, whilst likely to be seen as credible is compromised by mass produced bathroom scales which claim to produce the same value. In our work with a professional boxer, we reported a case where he began to believe he could compete at a lower weight following a discussion with highly successful boxing coach (1). In that case study, we describe how the boxer began to believe that he could make the lower weight and the advantage in terms of strength over his opponents. Sports scientists have evidence to support their ideas but do not necessarily have credibility. When a boxing coach asks the question: "which boxer has followed this type of diet before"?

What does the sport scientist say other than explain why it should work, and why following a diet developed for one person is also not always advisable. In the case study presented above, it is evident that care is taken to build relationships between key personal. The quality of the intervention is likely to be reflected in the quality of the relationship between coach, boxer and sports scientist (see (2, 3) for a discussion). The case study by James Morton and Graeme Close presents a compelling case for using sports scientists in the preparation of boxers.

References

1. Lane AM. Consultancy in the ring: Psychological support to a world champion professional boxer. In B. Hemmings & T. Holder (Eds). *Applied Sport Psychology.* (pp 51-63). John Wiley & Sons Ltd, London; 2009.
2. Lane AM. Reflections of professional boxing consultancy. *Athletic Insight.* 2006;**3**(8). *http://www.athleticinsight.com/Vol8Iss3/Reflections.html*
3. Schinke RJ. The Contextual side of professional boxing: One consultant's experience. *Athletic Insight.* 2004;**6**(2). Retrieved October 2004, from http://www.athleticinsight.com/Vol6Iss2/Professionalboxing.htm

CHAPTER 21

Coordination Dynamics in Olympic Archers

Richard Shuttleworth
Australian Institute of Sport, Australia.

Vignette

Winning the Sydney Olympics in 2000 was a career highlight for the then gold medallist and now current Australian archery head coach. He is now faced with the challenge of emulating a similar type performance with his young and talented archers based at the Australian Institute of Sport. For the Australian men's recurve team a key performance indicator in preparation for the 2012 Olympic Games in London, UK was the recent 2010 Commonwealth Games held in Delhi, India, and it was here that they won the gold-medal match.

To identify any rate limiters to performance in preparation for Delhi and London Games, a multidisciplinary perspective would be needed. After initial consultation, it was decided a coordination profiling approach would be appropriate because each individual athlete will satisfy the many interacting constraints in unique ways (1). Coordination profiling analysis involved collecting time-series data of individual and team performance outcome scores, kinematic and kinetic force plate data and athlete-bow interaction (2). In using a dynamical systems approach to optimising skill acquisition (3), it is firstly important to identify any rate limiters before manipulating key constraints. The analysis lead to the identification of movement coordination, control and attentional-focus constraints as rate limiters resulting in the development of a constraints-led intervention to affect behavioural change (3).

Discussion

There is limited research investigating the coordinative nature of aiming type tasks. In 1968, Arutyunyan, Gurfinkel and Mirskii examined movement coordination and varying levels of functional variability was found in both skilled and unskilled pistol shooters. In this study the skilled shooters demonstrated higher levels of variance around the shoulder and elbow joints which contributed to their success. It has been suggested the reason skilled shooters exhibit differing levels of variability to the lesser skilled is their ability to coordinate multiple joints in a complimentary manner to allow for stability in the wrist. The strategy used by the lesser skilled shooters involves lower levels of upper limb joint variability, this rigidity in upper limb coordination affects the ability of the wrist and pistol to remain aligned with the target. Similar findings have been found in a study on elite archery, where small fluctuations, referred to by Halliday, Conway, Farmer, and Rosenberg (1999)(4), as "physiological tremor", have been found to be caused by activity of the neuromuscular system for adjustment of arm posture (5). Furthermore, variability has also been found in on-target trajectories in highly skilled archers which demonstrate the interceptive nature of timing a release to hit the centre of the target. A study conducted at the Australian Institute of Sport which sampled all finalists at an International event, scientists found that maximum sway speed around centre of pressure (COP), to be highly correlated with outcome scores. In this study, the less successful archer's recorded higher levels of anterior-posterior sway both prior to and during arrow release, compared to the more successful archers in the competition (see Figure. 1). However, the competition place getters all demonstrated increasing levels of maximum sway during clicker and arrow release. These findings support recent literature which demonstrates that elite athletes performing accuracy tasks for example basketball shooting, demonstrate increased level of movement variability in a distal joint prior to implement release (e.g., (6)).

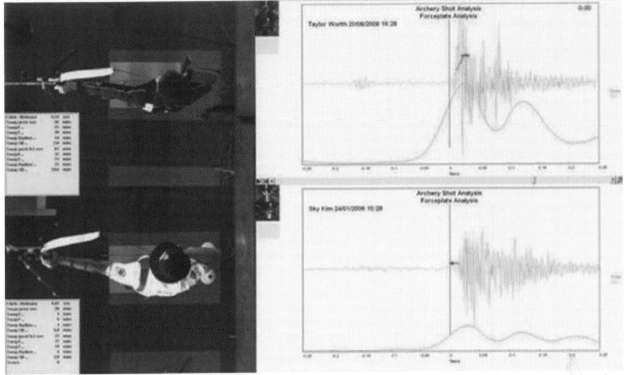

Figure 1. Overhead camera footage of two archers on force platforms competing in an International event held at AIS. The graph on the top right represents a less successful finalist and the bottom right graph represents a top finishing finalist.

The top graph represents an unsuccessful performer in the competition and the bottom graph represents a finalist. The solid red line at the bottom of each graph running along x-axis indicates maximum anterior-posterior sway. It is increasing in both conditions and continues to rise during the onset of the clicker (indicated by the vertical black line in the middle of graph), and during arrow release (indicated by the noisy light red line running along x-axis half way up graph). Arrow release was measured by an auditory recording device which was synchronised with force platform to provide accurate maximal sway speed with arrow release timing. Results from this study demonstrate that the successful competitors were able to shoot their arrows with a greater degree of control as indexed by lower levels of maximal sway speed. The increase in anterior-posterior sway may have been situated within their optimal bandwidth range for successful performance. The less successful performers struggled to maintain their form throughout the execution of the shot and this would suggest a coordination and/or physical rate limiter on performance. It has been suggested that adopting a rigid, hyper extended position with increased humeral/scapular muscle activation, may lead to a reduced degree of control over movement pattern trajectories and furthermore, increased instability in wrist-bow and eye-target alignment. Future research is needed to investigate these ideas. Finally, an overview of the speed accuracy trade-off in sport skills by Carlton, Chow & Shim, (2006)(7), suggest that training athletes for speed does not result in a loss of accuracy, at least not when participants perform under instructions for both speed and accuracy. This supporting evidence for placing athletes under time constraints while performing their skills may actually help improve an archers accuracy while reducing variability in timing with increased movement speed. Furthermore, recent feedback from archers who have practiced under time constraints have experienced reduced spatial-temporal inconsistencies in performing their technique with improved rhythm, tempo and confidence in reproducing skill.

Conclusion

As part of our individual case study analysis the first National training camp held at the AIS was used as a pre-test and the final selection camp was used as post-test. The AIS men's recurve athlete's (n=3), were subject to a constraints-led training intervention prior to the final Australian selection camp for 2010 Commonwealth Games. A three phase constraints-led intervention was designed by AIS skill acquisition specialist and the National head coach. The skill acquisition perspective taken is one of an adaptive relationship between the learner and their performance environment. It was our intention to facilitate this adaptive learning process by;

1. Destabilising an existing coordination pattern,
2. Assembling and refining a new coordination pattern
3. Acquiring adaptive movement under perturbation and within a team dynamic.

First, a destabilising period (phase 1); the aim was to perturb the existing movement patterns of the AIS athletes sufficiently enough to allow the system to switch states and eventually settle into a new more functional pattern of coordination. This phase was individualised to cater for the existing intrinsic dynamics of each athlete and therefore varied in time according to how stable each athletes previous technique was or whether their technique only needed slight adaptations to switch into a new pattern. Existing coordination patterns were very similar between AIS athletes due to the previous Korean model which emphasised high repeatability of an idealised technique. It is believed this approach may have promoted specific physical and performance limiters on a number of athletes'. A relatively short phase 1 period aimed to inject high levels of movement variability into the system. This involved each athlete undertaking a shooting practice schedule with key constraints manipulated for example; wearing a weighted vest which increased whole body sway forward, sideways and backwards, wearing a pair of Masai barefoot technology (MBT) shoes to induce forward-backward sway and using differing density foam surfaces underfoot to provide varying levels of instability into the lower body. The athletes became increasingly variable both in their existing movement patterns and their shooting scores. It was explained to each athlete beforehand what the likely affects on performance might be as a result of undertaking phase 1 of the skill learning intervention. Athlete confidence can be more sensitive during this early phase of movement adaptation with less consistency and control over outcome and process so this does need to be managed carefully.

Figure 2. AIS archer and Commonwealth Games gold medallist undertaking a constraints-led intervention. The vest provides varying weight distribution at front, side and rear of upper body depending on which aspect of technique will be perturbed. This strategy demonstrates how variability in system dynamics, exemplified by fluctuations in stability, should not be considered as system noise but as a functional property which allows the athlete to explore new states of order.

Phase 2 required each athlete assemble a new dynamic pattern of coordination which approximated and incorporated specific biomechanical movement principles believed to allow for greater control of movement execution. The new movement pattern involved several coordinative structures functioning as one dynamic whole motion (i.e., hips, trunk and shoulder rotation), with certain bandwidths of kinematic variance permissible for adaptive purposes both in relative sequencing and range of motion. DV cameras and Dartfish motion analysis software were used to qualitatively assess kinematic time series footage from front side, rear and top view angles. Varying informational constraints were used to help guide and shape the athletes coordination patterns. To avoid the often detrimental effects of adopting and internal focus of attention (8), and encourage a more implicit learning style for skill robustness under pressure (9), analogies and descriptors were used to create individualised topological landmarks to closer approximate a desirable movement pattern (e.g., think of your forearm movement as a sliding door on a van. Information constraints (i.e., instruction and feedback), were used to manipulate athletes attention toward an external-focus on movement effects (10), to enhance self-organisation and control of coordination (e.g., imagine piece of string pulling your elbow backwards, thereby resulting in forearm-arrow alignment). Blank face targets were used early on so that scoring or knowledge of results (KR), could not be used to constrain the ability of the learner to explore the task space fully (11). Delayed visual feedback was used via large flat-screen, DV cameras and Dartfish software. This enabled the use of self selected feedback and comparison between actual and athlete perceptions of their movement. Athletes' were also provided with summary feedback by the coach encouraging more active learning and emphasising movement pattern exploration and adaptation. The 'soft assembled' coordination pattern would afford the archer a release position which was more robust yet adaptable for individual's to shape to their own individual constraints (e.g., limb lengths and mobility), and importantly to adapt during the execution of the shot to various internal and external perturbations (e.g., anxiety, attentional effects, wind, time constraints etc,).

Figure 3. Biomechanical analysis of both traditional and alternative technique simulated by Simon Fairweather gold medallist at the 2000 Sydney Olympics and head coach of Australia's National team for 2012 London Olympics.

Phase 3 involved athletes acquiring adaptive movements. In Newell's (1985)(12) model of constraints on motor learning, task constraints are described as time dependent and consist of task goal, rules and regulations and equipment. It was decided to manipulate key constraints: time, competition rules, and perceptual variants to allow for the transfer of more adaptive movements into the performance setting. Athletes were scheduled to practice to pre-recorded vocal countdowns ranging between 12 seconds to 6 seconds. The countdown was varied each time hearing only the start and last 1, 2 or 3 second beeps leaving the athlete uncertain as to exactly how long they had remaining The aim was to attain spatio-temporal consistency in their shot whilst being subjected to temporal perturbations. Athletes performed as individuals and in groups of 3 to reflect team event. Several ends involved shooting down the centre of a soccer field, the perceptual flow gave athletes the impression that the marked 70m range suddenly

131

appeared 80m and the lack of reference points (e.g., lane ropes, fence lines), clearly provided additional visual perturbation. Modifications to the rules reduced significantly the number of arrows in each end thereby increasing the pressure to make every arrow count. This also included a self nominated double or nothing arrow which encouraged them to exploit with an accurate shot. Targets would be lined up in a row and pairs of shooters lined up. After each end the winner would move one place to the right and the loser would move to the left. The aim was for each athlete to climb the rankings by the end of the session and end up at the far right of the line. Rankings were explicitly known after each end by visually checking each athlete's position, this added additional psychological pressure. Surface was altered using grass, matting and concrete, injecting small amounts of variability into the lower limbs and was used in a randomised manner. On completion of each camp athletes were asked to peer review, provide group feedback and highlight challenges, problems and solutions which they learnt from. Results of the AIS men's recurve across pre-post test was an increase of 19 ± 7 and control 9 ± 11. The largest score increase was by an AIS athlete ranging from pre-test 315 to post-test 340. The intervention has provided learning evidence to feed back into a skill development pathway for sub-elite archers in the future.

References
1. Davids K, Glazier P, Araújo D, Bartlett RM. Movement systems as dynamical systems: The role of functional variability and its implications for sports medicine. *Sports Med,* 2003;**33**:245-260.
2. Spratford W, MacIntosh C, Davis M, James M, Turner R. The Impact of sway, equipment and release timing on scoring in elite level archery. 2009 In Mills P, editor. 7th Australasian Biomechanics Conference; 2009; Griffith University, Gold Coast, Australia; 2009. P. 79
3. Davids K, Button C, Bennett SJ. Dynamics of skill acquisition: A constraints-led perspective. Champaign, Ill. 2008: Human Kinetics.
4. Halliday DM, Conway BA, Farmer SF, Rosenberg JR. Load independent contributions from motor unit synchronisation to human physiological tremor. *J of Neurophysiology,* 1999;**82**: 664-675.
5. Lin JJ, Hung CJ, Yang CC, Chen HY, Chou FC, Lu TW. Activation and tremor of the shoulder muscles to the demands of an archery task. J of Sport Sci 2010;**28**(4):415-421.
6. Button C, Macleod M, Sanders R, Coleman S. Examining movement variability in the throwing action at different skill levels. *Research Quarterly for Exercise and Sport,* 2003;**74**(3):257-269
7. Carlton LG, Chow JW, Shim J. Variability in Motor Output and Olympic Performers. In Davids, K. Bennett, S. Newell (Eds.), *Movement system variability* (pp. 85-108). 2008 United States: Human Kinetics.
8. Wulf G, McNevin NH, Fuchs T, Ritter F, Toole T. Attentional focus in complex skill learning. *Research Quarterly for Exercise and Sport,* 2000;**71**(3):229-239
9. Masters R, Law J, Maxwell J. Implicit and explicit learning in interceptive actions. In K. Davids, G. Savelsbergh, S.J. Bennett, & J. Van der Kamp (Eds.), *Interceptive actions in sport: Information and movement* (pp. 126-143). 2002 London: Routledge
10. Wulf G, McConnel N, Gartner M, Schwarz A. Enhancing the learning of sports skills through external-focus feedback. *J of Motor Behaviour,* 2002;**34**(2):171-182
11. Muller H ,Sternad D. Decomposition of variability in the execution of goal-orientated tasks: Three components of skill improvements. *J of Experimental Psych: Human Perception and Performance,* 2004;**30**(1):213-233
12. Newell KM. Coordination, control and skill. In *Differing Perspectives in Motor Learning, Memory, and Control,* ed. D.Goodman, R.B.Wilberg, and I.M. Franks, 295-317. 1985 Amsterdam: Elsevier Science

Commentary: *Keith Davids,* Queensland University of Technology, Australia

This case study describes an intervention at the cutting edge of research exemplifying how learning design can be underpinned by principles of ecological dynamics (1). The methods employed in the AIS Skill Acquisition section eschew the acquisition of a common optimal pattern through a 'movement reproduction' strategy (2). Rather the focus is on skilled behaviour as adaptation to interacting constraints of each individual, specific task and the environment. The inspiration behind Newell's (1996)(3) constraints-led perspective on the soft-assembly of skilled movement solutions was Bernstein's notion of *dexterity.*(4) Newell's (1996)(3) proposed that "...dexterity, is an ability to solve a motor problem - correctly, quickly, rationally, and resourcefully. Dexterity is finding a motor solution for any situation and in any condition..."(p398). This aspect of expert movement behaviour is typically downplayed in traditional reproductive approaches to skill acquisition, but forms the theoretically principled basis for inducing fluctuations in the performance environments of the archers.

An initial step is to identify key rate limiters on the performance of each athlete and to consider manipulations to interacting constraints to drive the individual to discover and explore functional movement patterns to achieve performance outcomes. Coordination profiling is a technique emphasising how individual performers assemble a relatively unique performance solution in satisfying performance goals. The case study illustrates how inducing movement system variability may enhance stability in an archer's performance by creating information to regulate postural adaptive behaviours. Rather than attempting to eliminate postural sway, the case study showed how creating postural fluctuations can lead athletes to modulate standing balance to facilitate performance of actions nested on the fundamental task of standing upright and still (5).

The practical interventions in archery demonstrate how variability in system dynamics, exemplified by fluctuations in stability, should not be considered as system noise but as a functional property which allows the athlete to explore new states of order. Flexibility in adapting to localised performance conditions is enhanced by the capacity of the micro-components of a coordinative structure (muscles, joints, limb segments) to functionally vary in relation to each so that a performance outcome goal can be achieved. As noted in the case study, the principles behind coordinative structure theory were exemplified long ago in a study showing how the wrist and shoulder movements of skilled pistol shooters were functionally co-varied in order to perform an accurate shot on target (6). The emergent characteristic of skilled movement behaviour is highlighted as the dexterous adaptability of the performer to variable organismic, task and environmental constraints pressurising the topology of the movement pattern. The case study exemplifies how highly skilled athletes can precisely refine a coordination pattern to achieve a performance outcome goal, even in the most stable of performance conditions.

The role of the coach/skill acquisition specialist may be less prescriptive than previously believed. Manipulations of constraints on the archers revealed how novel, and more functional, movement solutions were assembled and tested. The role of coaches and skill acquisition specialists is to direct the search of performers towards more functional solutions.

References

1. Pinder R, Davids K, Renshaw I, Araújo D. Representative learning design: Implications for sport psychology. *J of Sport and Exer Psych*, 2011;**33**: 146-155.
2. Araújo D, Davids K. What exactly is *acquired* during skill acquisition?
3. *J of Consciousness Studies*, 2011;**18**:7-23.
4. Newell, K.M. Change in movement and skill: Learning, retention and transfer. In Dexterity and its Development (Edited by M.L. Latash & M.T.Turvey), pp. 393-430. 1996 Mahwah, N.J.: LEA.
5. Bernstein, NA. The Control and Regulation of Movements. 1967 London: Pergamon Press.
6. Riley MA, Stoffregen TA, Grocki MJ, Turvey MT. Postural stabilization for the control of touching. *Human Movement Science*. 1999;**18**: pp. 795–817.Arutyunyan GA, Gurfinkel VS, Mirskii ML. Investigation of aiming at a target. Biofizika 1968;**13**:536-538.

CHAPTER 22

Athlete Adaptation: A Comprehensive Approach to Intervention

Rob J Schinke, Randy Battochio & Katherine Johnstone
Department of Human Kinetics, Laurentian University, Sudbury, Ontario, Canada.

Vignette

I was approached by a boxing management group to assist an athlete with his performance preparation. I knew the boxer (pseudonym name James) from his amateur sport career, where I assisted onsite when he performed successfully at several major games. After turning professional boxer, he was touted as the future of the company that signed him. The first 15 bouts were relatively easy for James given his talent, well-established boxing skills, and also the quality of his opponents. James, when faced with higher quality opponents, started to realize that there were aspects lacking in his performance; either he was not as good as he initially believed, or he lacked the structure to sustain a successful professional career. When I started meeting with James he was in conflict with his coaching and management staff, and in fact he was at a contractual standoff. At the core of the conflicts was self-doubt – James anticipated that he was not going to meet the expectations of his fan base, and those he worked with.

Schinke met with the athlete numerous times and also with those in his staff. Over the course of seven months of discussion, an adaptation intervention was employed. The first objective was for the athlete to *understand* where his fears originated and why he was under-achieving in performances. After much discussion on the part of the athlete and the SPC it became clear that James was placing far too much pressure on himself through interviews with the media. In addition, there was insufficient understanding of each opponent well in advance of bouts. Examples included a limited appreciation of the opponents' tactics throughout the bout, such as being a fast or slow starter, a counter-puncher, and the distance the opponents preferred to box from. Instead, the athlete relied on his talent and also his capacity to adjust in the ring. Adding to the complexities, James lacked *trust* in his staff and he felt a lack of *belonging* within his team. Without trust, James lacked a belief in those able to assist him.

With an appreciation that performance preparation requires a personal strategy developed well in advance of the bout, James began using psychological skills as a key part of preparation. Within, he developed and posted his goals, which included personal skills in relation to his next opponent. Further, understanding extended to the importance of communicating openly with the coaching and management team, leading to shared understanding. From open lines of communication, James began to trust his team, and consequently, the team became highly collaborative. The subsequent bout was a step up in performance as James fought for a world title. In advance of the bout, James worked with his team and studied the opponent's tactics and his opponent's technical weaknesses. From a deepened understanding of what to expect, James managed himself well in pre-bout press conferences. Though he lost a unanimous decision, James continued to engage openly with his staff and also to employ a systematic adaptation approach to performance. The result two bouts later was that James became a professional world champion. The team continues to work together to present day and James has continued to advance as an athlete and also as a member of his larger team.

Adaptation

Adaptation within the vignette above best explains the process that transpired for James, leading to his expedited success as a professional boxer. Adaptation is a process comprised of five core motives the authors herein refer to as pathways. Built from the work of Fiske (1) and her precursor, Taylor (2), the pathways are comprised of *understanding, trust, belonging, control,* and *self-enhancement*. Within recent work proposed initially by Tenenbaum (3) and colleagues and then redefined by the first and second authors (4,5,6), there exist clear sub-strategies that were employed with James to build an adaptive approach to his boxing. Understanding entailed James' appreciation of himself. Through effective debriefings post-bout with the athlete individually and also with the team as a whole, correct and incorrect decisions were identified and subsequently documented 24 hours post-bout. In addition, James started using a diary twice per week, especially to seek understanding regarding the reasons behind variations in performance during training (7). Understanding also extended to an appreciation amongst James and his coaching staff regarding performance expectations, and how communication would be delivered during the bout. The understanding of the opponent was garnered through an opponent profile. The profile was built with attention to the opponent's disposition, pre- and post-bout attributions, tactics during press conferences and also how he entered the ring immediately before each bout. Also considered were the opponent's patterns throughout the bout, during each round and also in the corner after each round. Understanding also extended to an appreciation of the performance venue (8). Included within an understanding of the context were the exact dressing room used for the warm-up, how the entrance to the ring would unfold, the ring's size, lighting, and the referee's behaviors during previous bouts.

Trust also became a central part of the intervention strategy. James' trust in himself was garnered through detailed documentation of what works, acquired from debriefings and logbooks. As such, his self-trust was based on evidence, and not solely from his talent. Often talented athletes build their trust in themselves by solely relying on the talent that brought them to international level (9). What many forget is that at the international level, opponents often share the very same talent. Hence, talent is an important though in many cases neutral factor. Trust in the team for James was a big step forward. Arguably, it is easy for athletes to rely on themselves as they progress up the ranks, to the point where they begin to meet equally talented opponents. Once opponents are of equal caliber however, as bouts draw nearer in time, trust must extend to one's coaching and management teams. With James, trust was garnered through better – consistent communication with coaches and management. When there were differences in opinion, discussions ensued until problems were resolved and all were satisfied with the outcome. Of note, trust is much easier to maintain than to restore ongoing. Given that an athlete's current views are situated in relation to past experiences, the intent is to ensure that as many past experiences as possible affirm trust (10,11).

Belonging also became an essential part of James' overall intervention. James' team became a cohesive unit, socially and in terms of the task. The motto of the team has become "team is team", meaning that the membership stuck together no matter what. Included within the belonging pathway were belonging with the coaching staff and belonging with teammates. When an athlete perceives his coaching staff as collaborative and integrated, he also views his training and performance contexts as positive. Belonging with the coaching staff can be developed through team activities such as team jogs in the week before the bout, team meals, team meetings, and a unified front during press conferences and media interviews. Belonging also extends to a sense of belonging among teammates, meaning amongst athletes. When the team is crafted thoughtfully, the athlete's teammates should be comprised of others who are highly optimistic (10,11), equally motivated and minimally, equal in athletic credibility. When accomplished athletes train together as part of a team, they tend to push each other forward, with each athlete bringing out the best in those around him. Belonging with teammates becomes an essential part of the overall intervention in the final week pre-bout. The intent is for the athletes the client has trained with to support him during final tapering, during the weigh-in, in the dressing room immediately before the performance, during the walk to the ring, and immediately post-bout within the ring. Positive affiliations with synergistic coaches and teammates tend to build momentum and perspective, an aspect often lost by performers before they under-achieve in high profile opportunities (12,13).

Control delineates into confidence, assertiveness, and distraction control. Confidence can be built holistically as it was with James through understanding of self (i.e. increased self-awareness), the opponent, and the context. When understanding extends beyond personal strategies, especially in professional boxing when there is sufficient time to garner such understanding, the athlete becomes substantively confident. In addition, through the athlete's enhanced trust in his team, beyond himself, and also a synergistic training and performance environment built through belonging, the intent is to develop an amount of confidence that can match with the upcoming bout. Assertiveness manifests through better relations with coaching and managerial staff. With open communication, the athlete is able to express concerns, voice his opinions, and ensure that his views become a central part of the team's strategies and decisions. Distraction control within our team is often first addressed when we evaluate the performance environment and the opponent's tactics. Within the performance environment, one athlete I have met several times was taken off his focus when walking to the ring. Several people from the audience spit at him, which became a contextual intimidation factor. Though most distractions cannot be anticipated in advance, some can, thus decreasing what the athlete and his performance team are left to address onsite.

Finally, self-enhancement is comprised of effort and ability. With James, effort was enhanced through all of the pathways aforementioned. For example, through self-awareness, based on documented evidence from previous fights, James understood partially where to direct his efforts during daily training and also when nearing his next bout. An appreciation of the opponent's tactics and style also informed where James and his team directed their efforts (effort is in part contextually determined). In addition, through enhanced trust of his team and also a sense of belonging, effort was sustained daily through a positive working environment. Ability became more important as James started to compete against the best boxers in his division worldwide. Ability can be confirmed through video footage of previous successes (14,15), and also optimal footage taken during training sessions. When new aspects were learned or existing skills were refined, James' coaching staff ensured that developments were in part attributed to personal ability. In addition, video was used to identify tactical errors and minor technical oversights during sparring, with post-training coaching team discussions leading to refinements within each successive day.

Conclusions

Adaptation is a holistic approach to performance enhancement. Elsewhere (16), the authors have indicated that adaptation can superimpose on what is known from the sport coping literature (17). Matters of mastery and coping are central to our discussion whilst James is prepared for ideal and less than ideal circumstances. However, adaptation reflects a more detailed service model to prepare athletes such as James for the highest level of professional boxing. There are five pathways proposed to sport scientists and their clients, with each providing a viable entry point into the larger framework. Within each pathway, there are documented sub-strategies that SPC's can consider in relation to the client. Arguably the sub-strategies are re-packaged from mental training (12,18). Through adaptation however, the authors propose that mental training strategies can be considered thoughtfully as part of a theory informed approach to intervention.

References

1. Fiske ST. *Social beings: A core motives approach to social psychology* 2004 Danvers, MA: Wiley & Sons.
    2. Taylor SE. Adjustment to threatening events: A theory of cognitive adaptation. *Amer Psych* 1983,**38**:1161-1173.
    3. Tenenbaum G, Jones CM, Kitsantis A, Sachs DN, Berwick JP. Failure adaptation: An investigation of the stress response process in sport *Int J of Sport Psych* 2003;**34**:27-62.
    4. Battochio RC, Schinke RJ, Battochio D, Eys MA, Halliwell W, Tenenbaum G. The challenges of Canadian players in the National Hockey League. *J of Clin Sport Psych* 2009;**3**:267-285.
    5. Battochio RC, Schinke RJ, Battochio D, Halliwell W, Tenenbaum G. The adaptation process of National Hockey League players Submitted for publication.
    6. Schinke RJ, Battochio RC, Dubuc NG, Swords S, Apolloni G, Tenenbaum G. Understanding the adaptation strategies of Canadian Olympic athletes using archival data. *J of Clin Sport Psych* 2008;**2**:337-356.
    7. Devonport TJ. Perceptions of the contributions of psychology to success in elite kickboxing. *J of Sport Sci and Med* 2006;**5**(CSSI2):99-107.
    8. Schinke RJ, Ramsay M.World title boxing: From early beginnings to the first bell. *J of Sport Sci and Med* 2006;**8**(CSSI3):1-4.
    9. Schinke RJ. The relationship between support-infrastructure and athletic competence development in major-games. 2000. Unpublished doctoral thesis.
    10. Schinke RJ, Peterson C. Enhancing the hopes and performance of elite athletes through optimism skills *The J of Excellence* 2002;**6**:36-47.
    11. Seligman MEP. *Learned optimism: How to change your mind and your life,* 1991:NY: Pocket Books.
    12. Orlick T. *In Pursuit of excellence* 2008 (4th edition) Champaign, IL: Human Kinetics.
    13. Schinke RJ, Jerome W, Couture R. Social support and national team athletes with different perceptions *Avante* 2005;**11**:56-66.
    14. Lane AM. Reflections of professional boxing consultancy: A response to Schinke 2004. *Athletic Insight* 2006;**8**(3): Available from URL: http://wwwathleticinsightcom/ Vol8Iss3/Reflectionshtm
    15. Lane AM. Consulting in the ring: Psychological support to a world champion professional boxer In B Hemmings & T Holder (Eds), *Applied sport psychology* (pp 51-64) 2009, London, UK: John Wiley

16. Schinke RJ, Tenenbaum G, Lidor R, Battochio RC. Adaptation in action: The transition from research to Intervention Re-submitted to *The Sport Psychologist*

17. Kaissidis-Rodafinos A, Anshel, MH. Psychological predictors of coping responses among Greek basketball referees *The J of Social Psych* 2000;**140**:329-344.

18. Bull SJ, Albinson JG, Shambrook CJ. *The mental game plan: Getting psyched for sport* 1996. London, UK: Sports Dynamics

Commentary

William V. Massey & Barbara B. Meyer, University of Wisconsin-Milwaukee

Combat sports, and boxing specifically, are among the longest tenured sports in modern society. This fact, and the propensity for the media and lay public to highlight the barbaric nature of these activities, makes it somewhat surprising that there is a lack of research examining the biopsychosocial aspects of participation in these sports (1). While many outsiders regard boxing as an individual sport, Massey and Meyer (2) report that combatant athletes depend on coaches and training partners to enhance their physical and mental development as well as their performance. Our research and applied experiences provide support for the systems approach utilized by Schinke, Battochio, and Johnstone to improve James' preparation and psychological skills.

Schinke et al. describe an adaptation intervention in which the sport psychology consultant works with the athlete to develop understanding, trust, belonging, control, and self-enhancement, with each pathway contributing to the overall development and performance of the athlete. In the case of James, *understanding* was first approached by helping the athlete identify the origin of his fears. Understanding could also be approached by creating drills to help James simulate the demands of a competition, such as sparring sessions under bright lights with a rotating group of "fresh" training partners. This training drill would also require a high-level of *trust* in himself and his team, as well as a sense of *belonging* to the group. Thus, while Schinke et al. discuss developing an internal sense of trust, the consultant should also work with James to develop trust in his training partners, thereby enabling them to challenge him appropriately in sparring sessions. A sense of trust in self and others can also aide in facilitating a sense of control (3). For example, James can work with the consultant to create a "what-if" list for many of the uncontrollable and as yet un-experienced situations that have the potential to derail him and his team (4, 5). With a strong sense of trust in his teammates and training regimen, members of James' team can integrate items from the "what-if" list into his training. Additionally, by controlling his interactions with the media (e.g., blackout periods close to the fight, avoiding his own press), James can minimize not only the pressure of those contacts but also the pressure associated with a media-induced focus on winning and outcome. Finally, we concur that *self-enhancement* is gained through effort and ability, and thus strengthened through progress in the other four pathways. James' self-enhancement may be further augmented through viewing video highlights of his training and competition bouts.

In closing, we believe that the holistic and comprehensive method utilized by the authors of this chapter is consistent with the systems approach needed to provide psychological skills training to combat sport athletes. While additional scientific study is needed to more completely understand the needs of, and best practices for, this unique population, Schinke et al.'s work provides a good starting point for the examination of treatment paradigms in boxing and other combat sports (i.e., mixed martial arts). Future research should examine the role of coaches, training partners, and support staff in the development of peak performance among combat sport athletes. Only then can practitioners develop a deeper understanding of combat sport and the needs of the athletes within that particular sport context.

References

1. Wacquant L. *Body & Soul: Notebooks of an apprentice boxer.* 2004. New York: Oxford University Press

2. Massey WV, Meyer BB. An ethnographic study of the psychological factors in mixed martial arts. 2011, Unpublished raw data.

3. Bandura A. *Self-efficacy: The exercise of control,* 1997. New York: W.H. Freeman and Company.

4. Orlick T. *Psyching for sport,* 1986. Champaign, IL: Human Kinetics.

5. Vealey RS. *Coaching for the inner edge,* 2005. Morgantown, WV: Fitness Information Technology.

CHAPTER 23

The Realities of Working in Elite Sport: What You Didn't Learn in Graduate School

Barbara B. Meyer[1], Ashley Merkur[2], Kyle T. Ebersole[1], & William V. Massey[3]

[1]Department of Kinesiology, University of Wisconsin-Milwaukee, Milwaukee, WI, USA
[2] Olympic Winter Institute of Australia, Docklands, VIC, AUS
[3]Department of Occupational Therapy, Concordia University Wisconsin, Mequon, WI, USA

Vignette

One day before the start of on-snow training camp in Canada and four weeks before the first World Cup competition of the 2007-08 season, the veteran head coach of the Australian Aerial Ski Team (AAST) resigned. While the ability to identify and hire a high performance aerial ski coach is always a challenge due to the limited pool of candidates at the upper echelon, the short notice as well as the experience and talent levels of the team under study (e.g., 119 World Cup podiums; 5 World Championship medals; 7 Grand Prix titles; gold & bronze medals at the 2002 and 2006 Olympic Winter Games [OWG], respectively), made this a hugely demanding task. Four days later the Australians joined a large multinational team from Europe, led by a veteran head coach from Switzerland. A short-term solution for one season morphed into a long-term solution that, with ongoing modification in training venues and ancillary personnel, remained in place through the 2010 OWG. In the end, there were five athletes coached by four different individuals on two different continents, serviced by two primary support staff. See Figure 1 for a summary of team structure and location from November 2007 to February 2010.

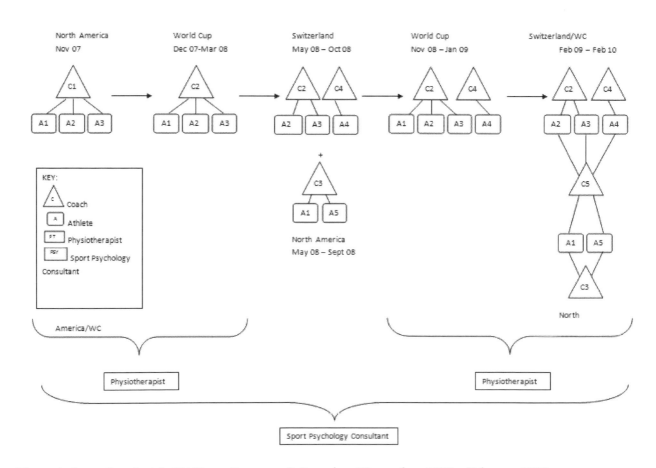

Figure 1: Australian Aerials Ski Team Structure & Location, November 2007 – February 2010

Informed by fluctuations in team structure and location, as well as intra-team competition, the work undertaken by the sport psychology consultant in the current case was more flexible and broad than is typical in Years 3 and 4 of an Olympic cycle (1,2). In the discussion to follow we describe the challenges faced by the sport psychology consultant in this atypically dynamic environment, and highlight how formal educational training should be augmented to prepare future professionals for this unique (but increasingly common) experience.

Discussion

Point #1 – Change in Training Location
The athletes on the AAST have maintained summer and winter training bases in North America for over a decade. By returning to the same regions of North America year after year the athletes acquire a level of cultural and practical familiarity that makes it easier to live and train abroad. The ability to plan and predict is important for athletes' perception of control, and indirectly for their sport performance and psychosocial development (3,4). This is particularly true in elite sport where athletes' schedules often change at the last minute due to uncontrollable factors such weather, travel complications, and/or injury. Although seasoned travelers and flexible yet disciplined individuals, the change in training venue from North America to Europe proved to be a major adjustment for many of the athletes in the current case.

From the perspective of the sport psychology consultant, more time and energy was spent working with the athletes on psychosocial issues than is normally done during Years 3 and 4 of the quadrennial. Specifically, time was devoted to helping the athletes manage:
(a) the language barrier associated with living in a country where English was not the native tongue;
(b) an unfavorable exchange rate and therefore the expense of groceries, particularly protein-rich foods (i.e., meat, fish);
(c) the sharing of coaches and facilities with athletes on a large European team, many of whom were their direct competitors, and;
(d) homesickness brought upon by the abrupt change in culture and a resultant sense of isolation.

While not insurmountable, the aforementioned challenges were amplified for athletes who were working to achieve and maintain a consistently high level of peak performance. For the sport psychology consultant, these novel and unforeseen circumstances meant that in addition to working on the typical mental skills training activities in the years and months leading up to the OWG, a great deal of time was also spent working with other staff members to minimize the impact on athlete performance of an ever-changing environment and lack of social support.

Point #2 – Teams within a Team
Despite originating from a country more revered for its beaches than its ski slopes, the AAST has been a prevailing force in World Cup skiing for over a decade. This success is not due to the achievements of one dominant athlete, but rather the consistent success of numerous athletes, who in the coactive sport of skiing, *compete against one another* despite training under the same flag. From the athletes' perspective, objective success may be fostered by the constant presence of one's competitors (5,6). From the staff perspective, there are numerous challenges to be negotiated in meeting the needs of individual members of the same *team* whose goals are mutually exclusive.

In the current case, the two most successful athletes on the team continued to separate themselves and their training programs as the Olympic quadrennial progressed. Moving into Year 3, the most senior athlete requested her own coach and the transfer of her training program back to North America. Moving into Year 4, the other athlete requested the addition of a coach who would serve as a technical specialist to the team. With the 2010 OWG less than 9 months away, the AAST Team maintained two different training sites. The athletes in Europe and North America were coached by two and one individual(s) permanently stationed at those respective sites, and three staff members (i.e., technical coach, physiotherapist, sport psychology consultant) who traveled between the sites.

From the perspective of all staff members, the commitment to provide each athlete the best possible program and therefore the best possible chance to succeed at the OWG took priority over the logistical challenges inherent in this intercontinental multi-team approach. Staff members were also sensitive to the

139

needs and perceptions of the three athletes on the team whose training programs were occasionally adapted to fit within the structure of the two marquee athletes (7,8). Although it may be atypical for team factors to be prioritized in a coactive sport, particularly in Years 3 and 4 of an Olympic cycle, the dynamic and tiered nature of the *team* structure in the current case prompted the sport psychology consultant to focus part of her work on managing team climate and cohesion. She consistently reinforced the credo that athletes did not have to like each other, but had to respect each other and each other's personal accomplishments. Similarly, that team members (i.e., athletes, staff members) were continuously moving in and out of training and competition environments meant that the sport psychology consultant was constantly monitoring stages of team development (9), and implementing strategies to facilitate a healthy and productive team climate for everyone. It should be noted that the team remained bifurcated through the 2010 OWG where one group of athletes and coaches operated out of the Olympic Village, another group of athletes and coaches operated out of a sub-site eight kilometers from the Olympic Village, and three staff members traveled between the two sites.

Point #3 – A Systems Approach to Service Delivery
Due in large part to the multifaceted and dynamic program constructed for the AAST, staff members spent much more time working directly with one another in the service of the athletes than is typical[2]. Prompted by current and previous experiences, the sport psychology consultant developed a model of managing athlete performance (see Figure 2), whereby ongoing collaborations among the various *elements* surrounding the athletes is paramount in facilitating their performance and well-being. This model fills a gap in the applied sport psychology literature, which acknowledges the importance of considering context when developing and implementing a treatment plan (e.g., Bronfenbrenner's Bioecological Model [10]; Systems Theory [11]; Carron's Framework for Cohesion in Sport and Exercise Groups [9]), but to date fails to acknowledge the importance of teams of professionals from different disciplines (i.e., physical, technical, mental) working together to treat clients. While the anecdotal observations informing this preliminary model have yet to be tested empirically, the efficacy of a team or systems approach to treatment has been supported in other disciplines (e.g., health behavior [12]; psychology [13]; medicine [14]), and holds promise in performance psychology as well.

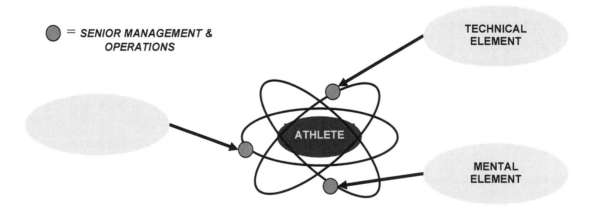

Figure 2: Meyer Athlete Performance Management Model

Conclusions
Athletes and staff involved in the aforementioned case are now able to look back and acknowledge the subjective (e.g., psychosocial lessons learned, interpersonal relationships developed) and objective (e.g., 17 World Cup podiums, bronze medal 2009 World Championships, Grand Prix titles 2008 & 2009, gold

---

[2] This was determined through discussions, with staff members and chapter authors, of previous professional experiences.

medal 2010 OWG) success accrued during that tumultuous 27-month period. Retrospective reflection does not change the fact, however, that staff often felt unprepared to manage the challenges brought about by this unique situation. For the sport psychology consultant specifically, little in the way of formal education or the research literature could be drawn upon to inform her work with this *team*. To that end, we offer the following suggestions for practitioners who work (or aspire to work) with athletes who compete at an elite international level.

- Be prepared to manage both performance-related concerns **and** psychosocial issues accompanying changes to longstanding protocols and predetermined plans. Refrain from assuming that seasoned travelers and veteran competitors will adapt with aplomb, and expect an increase in the frequency of contact from athletes/staff working through a dynamic environment.

- Anticipate that each athlete on a coactive team may require their own training regimen, including different support staff, training venues, and schedules. Commit to providing each athlete on the team the best possible program for achieving their goals. To that end, knowledge of social psychology and group dynamics theories along with experience in group and/or family systems counseling will enable the consultant to negotiate team issues that are sure to arise.

- Familiarize yourself with best practices literature on systems approaches to treatment, and commit to working **with** professionals in other disciplines to maximize the psychosocial development and sport performance of elite athletes. Be prepared to prioritize the safety, satisfaction, and success of athletes over your own personal schedule.

References

1. Blumenstein B, Lidor R. The road to the Olympic Games: A four-year psychological preparation program. *Ath Insight* 2007;**9**:15-28.
2. Vernacchia R, Henschen K. The challenges of consulting with track and field athletes at the Olympic Games. *Int J of Sport and Exer Psych* 2008;**6**:254-266.
3. Miller P, Kerr G. Conceptualizing excellence: Past, present, and future. *J of App Sport Psych,* 2002;**14**(3):140-153.
4. Taylor J, Wilson C. Intensity regulation in sport performance. In J.L. Van Raalte & B.W. Brewer (Eds.), *Exploring sport and exercise psychology* 2002 (2th ed., pp 99-130). American Psychological Association, Washington, D.C.
5. MacNamara A, Button A, Collins D. The role of psychological characteristics in facilitating the pathway to elite performance part 1: Identifying mental skills and behaviors. *The Sport Psych,* 2010;**24**:52-73.
6. MacNamara A, Button A, Collins D. The role of psychological characteristics in facilitating the pathway to elite performance part 2: Examining environmental and stage-related difference in skills and behaviors. *The Sport Psych,*2010,**24**.74-96.
7. Eys, M, Burke, S, Carron A, Dennis P. The sport team as an effective group. In J.M. Williams (Ed.), *Applied sport Psych: Personal growth to peak performance* 2010 (6th ed., pp. 132-148). McGraw Hill Higher Education, New York, NY.
8. McCann S. At the Olympics, everything is a performance issue. *Int J of Sport and Exer Psych,* 2008,**6**:267-276.
9. Carron A, Hausenblas H, Eys M. *Group dynamics in sport.* 2005 Fitness Information Technology, Morgantown,WV.
10. Bronfenbrenner U. Development ecology through space and time: A future perspective. In P. Moen, GH. Elder, K Luscher (Eds.), 1995 *Examining lives in context* (pp. 619-649). Washington, DC: APA.
11. Barker R, Garlock R. *Ecological Psychology: Concepts and methods for studying the environment of human behavior.* 1968 Stanford, CA: Stanford University Press.
12. Richard L, Gauvin L, Raine, K. Ecological models revisited: Their uses and evolution in health promotion over two decades. *Annu. Rev. Public Health,* 2010;**32**:307-26.
13. Malone D, Marriott S, Newton-Howes G, Simmonds S, Tyrer P. Community mental health teams (CMHTs) for people with sever mental illnesses and disordered personality. *Cochrane*

Database of Systematic Reviews, February 14, 2010;(3)Available from: Cochrane Database of Systematic Reviews.

14. Medves J, Godfrey C, Turner C, Paterson M, Harrison M, MacKenzie L, Durando P. Systematic review of practical guideline dissemination and implementation strategies for healthcare teams and team-based practice. *Int J of Evidence Based Healthcare,* 2010;**8**:79-89.

Commentary: *Rob Schinke*, Canada

Elite sport is often a very messy and complex environment for the applied sport scientist. When I began my doctoral studies, one of my professors shuddered when I conveyed my interest in professional practice. The response from that professor reflected her acknowledgement of the messiness of applied work. Sport psychology courses and textbooks often do not prepare the practitioner for the realities that they eventually face, once stepping out in the applied sector. The reality is that interventions must happen *in situ*, where understanding the sport context is the very baseline of effective intervention. Meyer, Merkur, Ebersole, and Massey are correct that understanding one's context and then developing one's work in relation to that context is perhaps the core of any successful intervention.

Effective practice must also be built collaboratively, where the sport psychology consultant ought to work in tandem with sport physiotherapists, nutritionists, coaches, and administration. An integrated approach to applied service allows the sport psychology consultant to piece her/his work together with the work of other staff in a manner that makes sense as part of a comprehensive approach. Conversely, there are applied sport psychologists who work in isolation and feel that their service should be a standalone offering to the athlete. Sport psychology consultants work within a social context where they must belong and also feel a sense of belonging to the team. From a collaborative approach, they can in turn earn the trust needed to provide effective service. Contextual understanding for those who work within a national team is essential, and that understanding can in part be gleaned by personal observations and experiences, though also indirectly from others immersed and committed to the very same national team context.

The reality for both athletes and staff alike is that extensive travel is often involved as part of the national team experience. Recently, Schinke, Tenenbaum, Lidor and Battochio (2010) have proposed an intervention model, built to facilitate elite athlete adaptation. National team athletes, though also professional athletes, encounter many typical competitive (i.e., tough opponents) and organizational (i.e., extensive travel) stressors as part of their athletic experience. With travel, comes several additional stressors, including the need to learn foreign customs, understand people who speak another language, and also an unfamiliar diet. Tallied together, these stressors can lead to stress overload on the part of the athlete. Consequently, the authors again are quite right to identify unfamiliar circumstances encountered through travel as "stressors". However, they might also seek to consider why unfamiliar circumstances sometimes become overwhelming stressors in the first place. Elite sport demands happen in a dynamic context where athletes must adapt to a multiplicity of unfamiliar circumstances. When a number of these demands happen simultaneously in contexts where the athlete must train and prepare for elite level performance, their compounded influence ends in a sense of being overwhelmed.

Finally, whether staff and consultants wish to acknowledge it or not, there are always teams within teams. Even when the entire team is under one roof, and committed to one set of coaches and service providers, teams tend to be comprised of smaller sub-teams, and even factions. Athletes of like mind are always drawn together, and the most successful athletes within such teams tend to often become the focal point for support staff, likely because they bring in government funding and notoriety that benefits all team members, including those in development and also, those past their prime. So, what to do if one wishes to bring these sub-teams together? The authors suggest that all team members look beyond their differences and adhere to a level of respect of each other. Correct, respect is one required and very basic mandate that all members in any group must commit to. I would think, that staff can work with their athletes and also embrace the differences among the team, whenever appropriate. Differences might be in terms of cultural practices, interpretations of a shared challenge, and so forth. Inviting athletes and staff to engage in shared problem solving or an exchange of perspectives also then brings a group comprised of sub-groups, closer together.

In closing, I agree full heartedly with the authors of this chapter that their work happens in a context that is messy. Courses could not possibly prepare the aspiring sport psychology consultant for such "messiness". However, I wonder if there is such a thing as a tidy social context? Perhaps course professors ought to look beyond their skills and tools and speak much more about social contexts, especially social contexts where the pressures associated with sport performance are the focus. Doing so will allow both teacher and students to then problem-solve, using a more holistic team approach, as opposed to sport psychology service, offered up in a vacuum.

Reference

1. Schinke, R. J., Tenenbaum, G., Lidor, R., & Battochio, R. C. Adaptation in action: The transition from research to intervention. *The Sport Psych,*2010;**24**: 542-557.

CHAPTER 24
Training to 'Draw-and-Pass' in Elite Rugby League: A Case Study.

Tim J. Gabbett [1,2], *Jamie M. Poolton* [3], *& Rich S. W. Masters* [3,4]

[1] School of Exercise Science, Australian Catholic University, Brisbane, Australia
[2] School of Human Movement Studies, The University of Queensland, Brisbane, Australia
[3] Institute of Human Performance, The University of Hong Kong, China
[4] Department of Sport and Leisure Studies, University of Waikato, New Zealand

Vignette

The ability of a ball carrier to draw in, or commit, an opponent to a tackle before passing to a teammate (i.e., 'draw-and-pass') in a 2-on-1 or 3-on-2 situation is fundamental to success in rugby league. Observational data show that of 303 tries scored by an Australian elite National Rugby League (NRL) team, 150 (49.5%) occurred as a result of a draw-and-pass (1). Of the tries scored, approximately 25% involved a series of two or more draw-and-pass attempts in the phase of play leading to the try. Furthermore, in the games analysed, 72% of successful draw-and-pass attempts and 47% of unsuccessful draw-and-pass attempts were simple 2-on-1 opportunities.

Being skilled at the draw-and-pass is the ability to achieve the intended outcome repeatedly, with little physical and mental effort and with appropriate speed and timing. In rugby league, acquiring this skill is a matter of effective movement (the best way in which to move in order to draw-and-pass) and efficient movement cognition (making the best decisions about where to move and when). The case study that we describe represents a theory–based approach to the training of improved draw-and-pass skills in a select group of players from an elite NRL team.

Background

As in most sports skills, becoming skilled at draw-and-pass is a matter of repetition (via practice), and is represented by consistency of performance and the ability to apply the skill adaptively in dynamic environments. Repetition is thought to allow internal representations of movement to be adapted and refined (using knowledge of results, for example) and to progress from information processing that is consciously controlled, slow and effortful to information processing that is unconsciously controlled, or automatic, fast and effortless. This stages of learning approach is central to most psychological theories of skill learning and underlies typical coaching methodology; however, Masters, Maxwell, Poolton and colleagues (e.g., 2-8) propose that conscious control is unnecessary even in the early stages of skill learning and performers may benefit from suppressing the involvement of conscious control in learning.

An *implicit motor learning* approach to skill acquisition has been proposed, in which interventions are imposed in order to prevent performers from consciously building up explicit knowledge and rules during learning (2). In essence, implicit motor learning interventions are designed to prevent learners from consciously processing and testing hypotheses about the relationship between their movements and the outcome of their movements. For example, interventions that restrict errors early in skill acquisition persuade learners that they do not need to test hypotheses about their movements because they are highly effective; consequently, learning is implicit (5). Removal of feedback during learning works in a similar way - a learner cannot test many hypotheses if no information is available about the outcome of the movement (6). Since Masters original work an implicit motor learning approach has been used to train performers with skills that remain stable in the face of psychological pressure (7), multi-tasking (5,9) and physiological fatigue (10-11).

In this particular case study, the coach chose to use a dual-task implicit learning approach. The dual-task approach is perhaps the most widely known and validated approach to implicit motor learning (e.g., 2, 12-13) and requires a learner to carry out a cognitive task in tandem with the skill that is being learned by repetition. The processing demands are so high that most performers are simply unable to think about their movements and thus do not consciously build up explicit rules and knowledge. Although, the dual-task approach has drawbacks – learning tends to be slow and effortful compared to normal approaches –

retention tests show advantages for later performance, including robust skills under competitive pressure and, importantly, an improved ability to multi-task.

The critical role that the draw-and-pass plays in rugby league has led to a program of research that has quantified draw-and-pass proficiency and examined the attention demands of the task for elite and sub-elite players (14). Dual-task paradigms are commonly used to evaluate the attention demands of a motor task by assessing the impact that a secondary (task-irrelevant) cognitive task has on performance (15). Dual-task draw and pass methodology discriminates professional NRL players from their semi-professional counterparts (16). High-skilled player's draw-and-pass proficiency is typically resilient to dual-task conditions, suggesting that successful execution of the skill relies on minimal attention demands. In contrast, lesser-skilled player's draw-and-pass proficiency degrades considerably in dual-task conditions (14), suggesting that for this group successful execution of the skill has higher attention demands. Interestingly, off-field dual-task draw-and-pass performance is predictive of the number of line break assists and try assists completed in a match (17) and is positively associated with on-field draw and pass success in 2-on-1 ($r^2 = 0.44$), 3-on-2 ($r^2 = 0.28$), and 4-on-3 ($r^2 = 0.23$) situations (1). This data implies that the ability of a player to proficiently execute a draw-and-pass in dual-task conditions is a characteristic of skilled performance that may significantly contribute to on-field rugby league success.

Training intervention

Players from the elite NRL team completed a simple 2-on-1 drill in a single task condition (4 trials) and a dual-task condition (4 trials), in which they simultaneously performed a tone recognition task that required them to verbalize a response to either a "high", "mid", or "low" pitched tone, by rapidly and accurately naming the tone. Verbal reaction time and response accuracy were recorded on each occasion. Draw-and-pass proficiency was quantified by a rater who employed a previously validated technical criteria checklist (14) (see Table I). On the basis of these draw-and-pass proficiency scores, we identified players who showed greater than 10% decrement in draw-and-pass proficiency in the dual-task condition when compared to proficiency in the single-task condition (N = 7). These seven players became the primary target group of a field-based, implicit motor learning training intervention.

Table I. *Technical criteria employed to assess the draw and pass proficiency of rugby league players.*

1.	Pass on the inside shoulder of the defender
2.	Small step away from the defender
3.	Body position square with defender
4.	Pass in opposite direction to leading leg
5.	Correctly identify when to pass and when to run
6.	Appropriate distance from the defender to prevent intercept pass
7.	"Take the defender away", "move the defender", "create a gap", or "open a hole"

The technical criteria were developed by two expert coaches and were consistent with cues used when coaching the skill of 'drawing and passing'. Players were awarded one point on each occasion that they achieved the relevant criteria and a score of zero if they failed to achieve the criteria. A total score (out of 7) was awarded for each of the trials. In addition, a total draw and pass proficiency score (expressed as a percentage) was awarded based on the aggregate of all technical criteria.

Players performed 10 implicit motor learning training intervention sessions over a 4 week period. Training drills were approximately 10 minutes in duration. Over the 4 week period, task complexity gradually increased. Initial training sessions involved simple 2-on-1 drills and then 2-on-1 games with points allocated for successful attack and defence. Later sessions involved 2-on-1 drills completed under variable conditions (e.g., ball placed in random starting positions, reduced field dimensions). The final training sessions involved simple 3-on-2 drills and then 3-on-2 games with points allocated for successful attack and defence. At all times, players performed the draw-and-pass activities while concurrently performing an irrelevant and non-specific secondary task. For example, players were asked to count backwards in 3's from 99, or forwards in 4's from 40. It was quickly recognized that the amount of player engagement in dual-task training was dependent on that player's relative interest in the secondary tasks. That is, it appeared that the success of dual-task training was reliant on the coaches identifying an appropriate secondary task for players on an individual basis. For example, arithmetic manipulation was suitable for the more 'academically-minded' players, but not for others. Some players responded well to listing as many types of car as possible while performing the draw-and-pass activities, while others were better at listing adult film stars!

Following the 10-session training intervention, players once again completed a simple 2-on-1 drill in a single task condition (4 trials) and a dual-task condition (4 trials). The two tests were completed for a final time following a 4-week non-training intervention retention period.

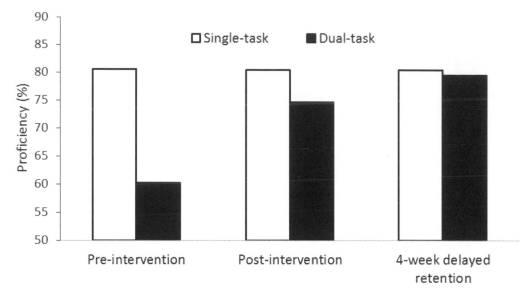

Figure I. Draw-and-pass proficiency scores of seven elite rugby league players in a single-task and a dual-task condition pre- and post-intervention and following a 4-week non-intervention training period.

The proficiency score data presented in Figure I suggests that the training intervention did not increase the single-task draw-and-pass proficiency of the players. Importantly, however, the data does imply that the training intervention caused a marked improvement in the player's capability to sustain draw-and-pass proficiency while attending to a secondary task (dual-task proficiency), suggesting that the draw-and-pass skill became less attention demanding. What is more, the robustness of player's draw-and-pass skill under dual-task conditions was consolidated over a 4-week period without prescribed dual-task training.

Our implicit motor learning training intervention successfully provided a sub-group of 'elite' player's with the dual-task capability to draw-and-pass, which is typically associated with skilled performance and is predictive of on-field success. Players competing in the NRL competition are faced with unique challenges. They are required to retain individual and team skills over prolonged periods of time (i.e., over a 26 match season), effectively execute these skills under high levels of pressure (e.g., playing in front of 50,000+ people) and physical exertion (see 18-19), while often processing multiple sources of information. It remains to be seen if incorporating implicit motor learning interventions into elite training programs allows robust performance in the face of such challenges, as has previously been observed with more novice performers.

References

1. Gabbett TJ, Abernethy B. Dual-task assessment of a sporting skill: influence of task complexity and relationship with competitive performances. *J Sport Sci* 2012;**30**:1735-1745.
2. Masters RSW. Knowledge, knerves and know-how. The role of explicit versus implicit knowledge in the breakdown of a complex motor skill under pressure. *Brit J Psychol* 1992;**83**:343-358.
3. Masters RSW, Maxwell JP. Implicit motor learning, reinvestment and movement disruption: What you don't know won't hurt you? In AM Williams and NJ Hodges (Eds), *Skill acquisition in sport: Research, theory and practice* (pp 207-228) 2004: London: Routledge.

4. Masters RSW, Poolton J. Unconscious learners, thoughtless performers: Developments in implicit motor learning. In AM Williams and NJ Hodges (Eds), *Skill Acquisition in Sport: Research, Theory and Practice* (2nd ed) (pp 59-75) 2012; London: Routledge.

5. Maxwell JP, Masters RSW, Kerr E, Weedon E. The implicit benefit of learning without errors *Q J Exp Psychol* 2001;**54A**: 1049-1068.

6. Maxwell JP, Masters RSW, Eves FF. The role of working memory in motor learning and performance *Conscious Cogn* 2003;**12**;376-402.

7. Liao CM, Masters RSW. Analogy learning: A means to implicit motor learning. *J Sport Sci* 2001;**19**: 307-319.

8. Orrell AJ, Masters RSW, Eves FF. Implicit motor learning of a balancing task. *Gait Posture* 2006;**23**:9-16.

9. Masters RSW, Lo CY, Maxwell JP, Patil NG. Implicit motor learning in surgery: Implications for multi-tasking. *Surgery* 2008;**143**:140-145.

10. Poolton JM, Masters RSW, Maxwell JP. Passing thoughts on the evolutionary stability of implicit motor behaviour: Performance retention under physiological fatigue. *Conscious Cogn* 2007; **16**:456–468.

11. Masters RSW, Poolton JM, Maxwell JP. Stable implicit motor processes despite aerobic locomotor fatigue. *Conscious& Cogn* 2008;**17**:335-338.

12. Hardy L, Mullen R, Jones G. Knowledge and conscious control of motor actions under stress. *Brit J Psychol* 1996;**87**:621-636.

13. MacMahon KMA, Masters RSW. Implicit motor learning: A suppression solution? *Int J Sport Psychol* 2002;**33**:307-324.

14. Gabbett T, Wake M, Abernethy, B. Use of dual-task methodology for skill assessment and development: examples from rugby league. *J Sport Sci* 2011;**29**:7-18.

15. Huang HJ, Mercer VS. Dual-task methodology: applications in studies of cognitive and motor performance in adults and children *Pediatr Phys Ther* 2001;**13**:133-140.

16. Gabbett TJ, Jenkins DG, Abernethy B. Relative importance of physiological, anthropometric, and skill qualities to team selection in professional rugby league. *J Sport Sci* 2011;**29**:1453-1461.

17. Gabbett TJ, Jenkins DG, Abernethy B. Relationship between physiological, anthropometric, and skill qualities and playing performance in professional rugby league players. *J Sports Sci* 2011;**29**:1655-1664.

18. Gabbett TJ, Jenkins DG, Abernethy B. Physical demands of professional rugby league training and competition using microtechnology. *J Sci Med Sport*, 2012;**15**:80-86.

19. McLellan P, Lovell DI, Gass GC. Performance analysis of elite rugby league match play using global positioning systems. *J Strength Cond Res* 2011;**25**:1703-1710.

Commentary: *Damian Farrow*, Victoria University and Performance Research, Australian Institute of Sport

There are a number of positive messages that I extracted from the Gabbett, Poolton and Masters case study. While skill acquisition research has a relatively long and rich history, systematic work of an applied nature is still in its infancy. What the authors were able to demonstrate was an example of how such work can be integrated into a high performance setting where sometimes the focus on the W or L column (win or loss) is all consuming

Of particular note was the evidence-based and systematic nature of the skill development intervention. First, a critical skill (draw-and-pass) was identified where individual improvement has the potential to make a significant contribution to overall team performance. Second, despite being a team environment, individuals who were most likely to benefit from additional skill practice were targeted through objective, context specific testing rather than the adoption of a one-size fits all approach. Third, sufficient time and a progressively challenging practice intervention were established within the players' programs such that an improvement in skill level could be expected. While each of these points seems rudimentary, experience suggests that attention to detail is often a missing ingredient in supposedly high performance sports settings.

Particularly refreshing was the application of an implicit learning approach, which has typically only been considered in novice populations (See[1] for an exception.) Implicit learning is often considered too experimental or risky by a high performance coach (albeit not exclusively) for a number of reasons. First, while the purported benefits of implicit learning such as enhanced pressure resistance and retention are intuitively appealing, the means through which it is created is often met with skepticism. To paraphrase a number of coaches I have discussed the concept with; "dual-tasking seems like a game you play with primary school children – rhyming words, solving simple mathematics." Yet as Gabbett et al highlight, not only is such a task challenging but finding the right level to pitch this dual-task is critical to its ongoing success. Furthermore, coaches are also concerned about the length of time such approaches take to develop skill. The mantra of needing to fast-track skill development is often used in response to suggestions of an implicit learning approach. Again, this aspect of implicit learning is acknowledged by the authors, and like them I think it's worth the investment. The authors post-test, and in particular retention test data provides further evidence of the value of such approaches and has the potential to convince more coaches of the value of implicit motor learning approaches in the high performance sports setting.

References

1. Rendell M, Farrow D, Masters RSW, Plummer N. Implicit practice for technique adaptation in expert performers. *Int J Sport Sci Coach* 2011;**6**:553-566.

CHAPTER 25

Working as a Physiologist in Professional Soccer

Barry Drust
The Football Exchange, Liverpool John Moores University, Liverpool, UK

Discussion

Soccer is characterised by high-intensity efforts superimposed on a background of endurance running. Both the aerobic and anaerobic energy systems (around 90 and 10% respectively; (1) are taxed during match-play. Soccer also requires players to perform specific match actions such as jumping, kicking, tackling and changing direction quickly that require the rapid development of force. Performance in such activities is related to the strength of the muscles used to produce movement. These general requirements of soccer are dependent to a large extent on the positional role of the player (2).

The participant in this case study was a male full-time youth trainee soccer player (age 17, height 1.82 m, body mass 78.4 kg) who played in midfield at a Nationwide League One football club. Midfield players typically cover the greatest distances during match-play (3) as a result of their role as a link between defence and attack. Increases in activity tend not to only reflect the greater distances completed at sub-maximal intensities (e.g. walking and jogging) (4) but may also reflect greater requirements for high speed running and sprinting (5). These specific positional requirements indicate that a high aerobic capacity as well as effective pathways for anaerobic energy provision is important for midfield players. Such fitness attributes will enable players to produce energy to develop and maintain a high-intensity of exercise for prolonged periods of time and to recover quickly after a bout of maximal exercise. The requirement to perform specific match actions will also necessitate an ability to generate force. The evaluation and development of a range of fitness parameters are therefore required to ensure that a midfield players' fitness matches the demands imposed on him during competition.

A needs-analysis was completed using a two-stage process. A performance profile, that included the key physiological requirements for the athlete's specific playing position, was presented to and completed by him in a private meeting. This process enabled me to obtain a personal perception of the clients perceived physiological strengths and weaknesses. This personal evaluation was followed by objective physiological tests. The use of physiological tests to provide an indication of their performance potential was preferred over the use of the individual's match performance. This was a direct consequence of the large variability observed in indicators of match performance (6). The inherent variability (which is a consequence of factors such as the specific tactical requirements of a given opposition, the score-line, the venue etc) of such data makes it unsuitable for the detection of the relatively small changes that may occur in physiological function as a consequence of the completion of a relatively short-term training programme. An emphasis on the interpretation of changes in the outcomes of carefully controlled physiological tests provides a more robust approach to detecting changes in performance status in interventions of this nature. Such data also allows the quantification of the individual's abilities in relation to group averages and other published data, as well as their own individual performance data, thereby enabling meaningful comparisons to be made with other individuals/populations.

The test battery comprised both laboratory tests and more soccer-specific field-based evaluations. The assessments were chosen based on their relevance to the fitness characteristics under consideration, their ease of use, their sport specificity and their time efficiency. Laboratory tests were used to provide a more general physiological assessment of aerobic fitness with field tests providing information relating more to soccer-specific performance. All field tests were performed on an artificial indoor surface where environmental conditions could be continuously controlled. All tests employed had been previously assessed for reliability using the standard error of measurement (SEM) (7). The player was fully familiarised with all tests prior to the relevant assessments. The following assessments were included in the test battery partly as a consequence of the analysis of the individual needs-analysis: Maximal Oxygen Consumption ($\dot{V}O_{2\,max}$); Yo-Yo Intermittent Recovery Test Level 2 (Yo-Yo IR2); Repeated Sprint Test (8).

The outcomes of the subjective aspect of the needs analysis suggested that the athlete's weaknesses were his "endurance" and his ability to "repeat high-intensity efforts". These views were supported by the objective data collected (see Figure 1). The player's $VO_{2\,max}$ value was lower than both the group average for other players in the squad and the mean values observed for other similar soccer players. Data from the Yo-Yo IR2 also indicated that the client's performance did not match that of other age matched individuals. The data from the repeated sprint test illustrated that while one off sprint performance was similar to other players the ability to perform repeated high-intensity bouts of exercise was inferior to others. This information would suggest that a major focus of any intervention should clearly be on the development of the athlete's aerobic fitness. Development in these areas would lead to improvements in the general aerobic fitness of the player but also increase the individual's capacity to support and recover from repeated intense bouts of exercise. Such physiological changes may enable performance of higher numbers of repeated bouts of intense exercise to be performed in matches and subsequently improve match activities.

A
MAXIMAL OXYGEN CONSUMPTION (ml.kg⁻¹.min⁻¹)

YOU	55.1
SQUAD AVERAGE	60.5
PLAYERS IN TOP LEAGUE	62.3

Comments
Score is lower than squad and other players. You need to develop this aspect of your fitness.

B
REPEATED SPRINT TEST

	YOU	SQUAD AVERAGE	PLAYERS IN TOP LEAGUE
BEST TIME (S)	**6.52**	6.57	6.61
MEAN TIME (S)	**6.95**	6.72	6.78
FATIGUE INDEX (S)	**0.68**	0.53	0.50

Comments
Your best score is good. The mean time and the fatigue index give us an idea of your ability to perform repeated sprints. When you compare your data with others this shows we need to work on this aspect of fitness.

C
YO-YO INTERMITTENT RECOVERY TEST (M)

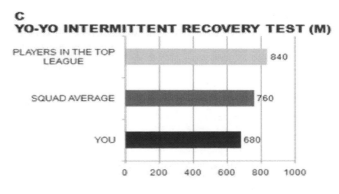

Comments
Your score here is lower than other players. This would suggest that you need to work on your ability to complete repeated high-intensity running bouts.

Figure 1: Results from the pre-test assessments for the individual player; (A) maximal oxygen consumption, (B) repeated sprint test, (C) YoYo IR2

150

The player followed an eight week training schedule. This training schedule was agreed with the individual's coach. The needs to continue the technical and tactical development of the player prevented him from been withdrawn from the squad to complete dedicated fitness work. The intervention was therefore designed around 3 supplementary training sessions. It is likely that such models of intervention are the most applicable to this specific sport as it is unlikely that alterations to the overall training programme completed by players are possible in the majority of clubs. The training intervention was predominantly aimed at improving the aerobic fitness of the player.

Training sessions were predominantly high-intensity in nature. This strategy was partly informed by the limited time that was available for additional training and the effectiveness of this type of training prescription for improving aerobic fitness (9) Short duration sprint based high-intensity training sessions were also included within the programme such short-term sprint interval training programmes have recently been shown to be a potent training stimulus for the aerobic energy system both generally (10) and in specific populations (11). The inclusion of this type of activity would also provide a specific anaerobic training stimulus for the client. This would be important as a basis for improvements in performance parameters that rely more on anaerobic energy provision such as discrete sprints or repeated high-intensity efforts over relatively short durations of time. Table 1 shows an example week from the training intervention

Table 1: Example week of supplementary training from the 8 week training programme

	Session 1	Session 2	Session 3
Session type	Soccer-specific interval session	Repeated sprint session	Soccer-specific interval session
Exercise prescription	5 x 4 min intervals at 95% max.heart rate	3 x 8 x 40 m sprints	2 x 10 x 1 min interval sessions at 95% max.heart rate
Recovery prescription	3 min at approximately 50% max.heart rate	20s between sprints, 4 min between sets	1 min recovery between each effort, 5 min between sets at approximately 50% max.heart rate

All of the training sessions were delivered using interval-training techniques. This provides a better representation of the intermittent activity pattern associated with soccer. The specificity of the training stimulus was further improved by using soccer-specific movement patterns during training as opposed to more conventional distance running techniques (i.e. track work or road running). All training sessions were carried out on a soccer field using relevant pitch markings to guide distances and to help guide the soccer-specific movement activity utilised. All recovery periods were active as this again replicates the activity profile that is observed in games. The client wore a HR monitor when possible during training sessions. He was also familiarised with Borg's RPE scale in order to rate the overall intensity of the session. This served to ensure that the correct intensity of exercise was achieved during the physical conditioning sessions.

Within two weeks of the completion of the intervention the pre-programme test battery was repeated (see Table 2). Improvements in the client's general aerobic fitness were observed following the intervention indicating that the players' general aerobic fitness was now in line with other players in his peer group and with the values observed for other midfield players. The total distance covered on the Yo-Yo IR2 was increased by around 180 m and relevant indices of the repeated sprint test (mean time, fatigue time) were also improved. This suggests that the player has improved both his ability to generate energy through the aerobic and anaerobic energy systems and/or improved the ability to recover from intense efforts. Such improvements may facilitate the completion of additional high-intensity running within games and hence improve the overall match performance of the player in question.

Table 2: Percentage change in performance test results for the individual player following the intervention

Performance Marker	% change from pre-test
Maximal Oxygen Consumption	9.8
YoYo IR2	26
Repeated Sprint test	
Best time	0.1
Mean time	3.6
Fatigue index	14.7

The successful attainment of the primary objectives of the support programme would seem to indicate that the case study represents a successful intervention though it is difficult to know if these discrete improvements will impact actual match performance. Some relevant considerations for similar future interventions may include (a) the inclusion of more dedicated anaerobic training sessions to facilitate greater improvements in anaerobic energy requirements, (b) potential changes in the sport-specific field based assessment protocols (i.e. the inclusion of specific discrete sprint tests at the expense of the repeated sprint protocol) and the way important parameters are calculated (e.g. the fatigue index), (c) an attempt to evaluate the impact of the training programme in a more soccer-specific performance context (that is not limited by the problems associated with using match-play activity profiles).

References

1. Bangsbo J. The physiology of soccer- with special reference to intense intermittent exercise. *Acta Physiologica Scandinavica*,1994;**151**:619.
2. Drust B, Atkinson G, Reilly T. Future perspectives in the evaluation of the physiological demands of soccer. *Sports Med*,2007;**37**:783-805.
3. Di Salvo V, Baron R, Tschan H, Calderon Montero FJ, Bachi N, Pigozzi F. Performance characteristics according to playing position in elite soccer. *Int J Sports Med*,2006;**27**,1-6.
4. Rienzi E, Drust B, Reilly T, Carter JEL, Martin A. Investigation of anthropometric and work-rate profiles of elite South American international soccer players. *J Sports Med and Physical Fitness*,2000;**40**:162-169.
5. Di Salvo V, Gregson W, Atkinson G, Tordoff P, Drust B. Analysis of high-intensity activity in elite soccer. *Int J Sports Med*,2008;**29**,1-8.
6. Gregson W, Drust B, Atkinson G, Salvo VD. Match-to-match variability of high-speed activities in premier league soccer. Int J Sports Med 2010;**31**:237-242.
7. Hopkins WG. Measures of reliability in Sports medicine and science. *Sports Med*, 2000;**30**:,1-16.
8. Bangsbo J. *Fitness Training in Football. A Scientific Approach*. 1994 Bagsvaerd, Denmark: HO+Storm.
9. Impellizzeri FM, Marcora SM, Castagna C, Reilly T, Sassi A, Iaia FM, Rampinini E. Physiological and performance effects of generic versus specific aerobic training in soccer players. *Int J Sports Med*, 2006; **27**:483-492.
10. Gibala MJ, McGee SL. Metabolic adaptations to short-term high-intensity interval training: A little pain for alot of gain? *Exercise and Sport Sci Reviews*,2008;**36**:58-63.
11. Ferrari Bravo D, Impellizzeri FM, Rampinini E, Castagna C, Bishop D, Wisloff U. Sprint vs. Interval training in football. *Int J Sports Med*, 2007;**29**:668-674.

Commentary; Ross Cloak, University of Wolverhampton, UK

The case study has raised some important aspects related to being a successful sports scientist working in an applied setting. Communication between the sports scientist, player and coaching staff being one of the most important.

The abundance of technology tracking player's exertion during match play means we are gaining a much clearer picture of what is required from each position on the pitch. This increases the validity of the initial needs-analysis in terms of demands on the player, and therefore, helps inform the subsequent test design and intervention strategy.

The initial private meeting with the player to discuss their perceptions of their physiological abilities can be a time consuming process with a large squad and sometimes neglected by many, who move straight to the physiological battery of validated tests. However, this can be an invaluable opportunity to meet with the player, without the perceptions of their coaches or peers and have open discourse of issues/weaknesses they face. It also provides the opportunity to build your own relationship with a player (who up to this point may have only seen you at limited points of the season and their experiences are one of the pain infliction and pocking and prodding!).

The next important line of communication is then with the coaching staff/management. Particularly important during high stress periods of the season where technical/tactical development takes priority (as well of course as the all-important 3 points). Drust puts forth a strong argument on how the improvement in repeated sprint ability will translate to match activities. This is a key point; the discussion with the coaching staff needs to make clear links between the additional workload and how it will benefit the player/team in terms that are easily understood. This may not allow "dedicated fitness work" as indicated, but it will help development a relationship of trust with the coaching staff that the additional work is worthwhile and therefore, more likely not to have its dedicated time in the training day "given away" to other aspects of the player's development.

Again, the time constraint the player has to perform any additional intervention helps to inform the sports scientist of the intervention that will provide maximum return (high intensity interval work is specific to nature of the sport, time efficient and easily quantified in terms of load on the player). As with a lot of team sports the idea of "additional" fitness work can be seen by many players as a punishment. It is important for not only adherence to the task, but also the moral of the player that this is not the case and the soccer specific movement patterns adopted in the case study, hopefully help keep the players interest and distract from the intensity of the session. As well, of course, as fitting into the initial needs analysis of the player's position and training goals.

A very relevant consideration raised in the case study is that of an attempt to evaluate the impact of a training intervention in a soccer-specific performance context. The idea would help add further weight to the importance of such interventions in the eyes of coaching and management staff who have a large say in time allocation available to the Sports Scientist.

CHAPTER 26
Physical Preparation of a World Champion Breaststroke Swimmer

David B. Pyne[1] and Vince Raleigh[2]
[1]Sports Science and Sports Medicine, Australian Institute of Sport, Canberra, Australia
[2]Swimming Program, Australian Institute of Sport, Canberra, Australia.

Vignette

A 25 year old Caucasian male swimmer, Brenton Rickard of Australia, won the 2009 FINA World Swimming Championship in the 100-m men's breaststroke in world record time of 58.58 seconds. Here we present an analysis of the long-term physical preparation and performance progressions of this athlete. We monitored progressions and variability in competitive performance, anthropometric measures and pool-based physiological testing over the six year period from 2003-2009 inclusive. Performance time to complete the 100-m breaststroke event in major events (national championships or international competition in a 50-m pool) was recorded electronically and the season best time established for each calendar year. Four compartment fractionation anthropometric testing was conducted approximately four times per year. A 5 x 200-m breaststroke step test was conducted regularly to monitor changes in maximal effort time trial performance, derived 4 mM lactate threshold and indicators of stroke efficiency. Rickard showed a 4.5% improvement in best time from 2003 to 2009 culminating in the gold medal and world record at the 2009 World Swimming Championships. Performance time improved by up to 1.6% from 2003 to 2007, then another 3.2% from 2007 to 2009: the latter margin reflecting in part the effect of rapid advances in swim suit technology. The world record in this event improved by 2% over the same period. Rickard's height did not change substantially from age 20-25 y although his body mass increased by 3.1%. Sum of seven skinfolds declined by 6.5%, while muscle mass (kg) and muscle mass (% of total body mass) increased by 2.7 kg and 1.4% respectively. There was a cyclical pattern of change in the 5 x 200 m step test performance measures within and between seasons. Maximal effort performance times, derived lactate threshold velocities and heart rates typically varied by only ~2%, yet blood lactate concentration varied by ~17%, and stroke characteristics by 5-8%. These patterns of change and variability in performance, anthropometric and physiological measures provide benchmark values for the long-term preparation of other breaststroke swimmers.

Discussion

The long-term preparation of a swimmer for success at the international level typically takes 10 years of training at a high level. This length of time is evident with male breaststroke swimmers who generally obtain their best performances in their mid 20's. A male swimmer's competitive career can generally be divided into junior (<17 y), transition (17-20 y) and senior (>20 y) phases. Here we report the pattern of improvements in anthropometric, physiological and performance characteristics in Brenton Rickard (Australia) the 2009 world champion 100-m breaststroke swimmer as a senior swimmer between the ages of 20-25 y. Breaststroke is a highly technical stroke and requires well-developed levels of swimming skills, fitness and strength for competitive success. Rickard's coach (VR) instituted a systematic sports science program for his squad involving regular pool and laboratory testing.

Brenton Rickard typically trained year round with most weeks consisting of nine or ten pool training sessions and three gym-based weight training sessions. The annual training plan was based around peaking for two main competitions: the national championships in March or April and the major international competition (e.g. World Championships or Olympic Games) in July or August each year. The early season training volume of 30-40 km.week[-1] was increased to ~50 km.week[-1] for the main training phase (4-10 weeks before competition) before a marked reduction (typically ~60-80%) in training volume (the taper) prior to the main competition. The taper is a key element of the physical preparation of athletes in the weeks immediately preceding competition (1). The importance of reducing training volume was highlighted in a meta-analysis (2), which indicated that performance improvement is more sensitive to reductions in training volume than manipulation of other training variables. The magnitude of improvement with reduced volume is approximately twice that of modifying training intensity or training frequency (2). We consider an important aspect of the swimmer's preparation was careful management of training, fatigue and recovery during the taper period prior to major competition.

A series of studies on Olympic, senior and junior swimmers established reference values for the typical rate of improvement within a competition (heat-semi-final) and over a competitive season (national championships to international competition) (3-5). A systematic improvement in competitive performance within- and between-seasons is required by an individual swimmer to substantially improve their chances of international success. Our initial modelling showed that to stay in contention for a medal, an Olympic swimmer should improve his or her performance by~1% within a competition and by ~1% within the year leading up to the Olympics (3). However these estimates were generated prior to the introduction of the fast skin swimsuits in early 2008. Rickard's improvement of 3.2% in 100 m breaststroke time in the two years prior to his world championship success was substantially greater than the change in the world record over the same period (Figure 1). In our experience, a swimmer ranked in the Top 10 in the world, and who can make a substantial improvement in performance time, has a good chance of a medal.

Rickard was considered to be a late developer in comparison to many of his peers in the sport. By age 20, he had developed a mature physique with a height of 1.95 m and a body mass of 90 kg (Table 1). Developing and maintaining lean body or fat-free mass is generally considered important in sports (like swimming) where speed, power, and strength are associated with performance (6). During the study period the swimmer maintained a lean powerful physique with a low sum of skinfolds (fat mass). Within a season Rickard's sum of skinfolds typically followed a similar pattern to other swimmers in showing a late season improvement prior to major competition (7). The coach's strategy was to focus on strength and resistance training during the early part of the season (increasing to 4 gym and 2 pilates sessions per week) before concentrating on swim training and racing during the second half of the season.

Table 1 Anthropometric changes in a world champion breaststroke swimmer over a five year period. SD = standard deviation, %CV = coefficient of variation, and muscle mass (% of total body mass

	Height(cm)	Mass(kg)	Skinfolds(mm)	Muscle Mass(kg)	Muscle Mass(%)
Feb 2005	194.6	90.0	49.4	44.3	47.5
Jul 2009	194.4	92.8	46.2	45.5	48.1
Mean	195.1	92.1	48.0	45.5	48.2
SD	0.3	2.1	3.4	0.9	0.3
%CV	0.1	2.3	7.2	2.0	0.7
%change	-0.1	3.1	-6.5	2.7	1.4

The 5 x 200 m swimming step test involves a series of 200 m swims performed in a progressive incremental manner. The test is usually completed in freestyle although for breaststroke specialists the test is swum in that stroke. The 5 x 200 m step test (and variations such as the 7 x 200 m step test) are used frequently as part of an elite swimmer's training program to monitor progress in fitness, prescription of training speeds, and prediction of competitive performance (8-10). The step test involves performance (times, split times), physiological measures (heart rate and blood lactate) and stroke characteristics (stroke count, rate and efficiency). A combination of step-test measures can be used to predict mean and individual changes in competition performance (10) including breaststroke swimming (11). This world champion swimmer exhibited a high level of consistency in step test performance (Table 2). The mean variability in step test performance times was low at ~2%, with more variability in stroke rate, count and efficiency (5-8%) and the highest variability in blood lactate concentration (17%). A cyclical pattern of improvement in training performance within and between competitive seasons (data not shown) is consistent with an earlier study from our laboratory (8). Individual test results should be interpreted in light of recent training history and time in the season. Coaches should emphasise all of the kinematic components in training and identify the stroke rate to stroke length ratio most appropriate for the individual swimmer (12). Measuring stroke characteristics in routine training and testing is an important element of improving the arm-leg coordination in breaststroke swimming (13-15).

Table 2. Descriptive analysis of 17 different 5 x 200 m step tests undertaken a 5 year period leading to a world record. SD = standard deviation, CV = coefficient of variation; min = minimum value, max = maximum value, s = seconds, stk = strokes; # = number of strokes, SEI units = velocity (m.s⁻¹) x stroke length (m).

	mean	SD	%CV	min	max
1st 100 m split (s)	72.8	1.5	2.1	70.7	75.2
2nd 100 m split (s)	76.1	1.7	2.2	74.1	79.7
200 m time (min:s)	2:28.97	0:03.12	2.1	2:24.80	2:34.38
Heart rate (bpm)	167	4	2.5	160	176
Lactate (mM)	4.7	0.8	16.7	3.2	5.8
3 mM time (s/100 m)	77.4	0.9	1.2	75.9	79.5
4 mM time (s/100 m)	75.4	1.1	1.5	73.7	77.7
8 mM time (s/100 m)	70.9	2.1	2.9	65.8	75.4
Stroke rate (stk/min)	29.7	2.5	8.3	25.8	35.5
Stroke count (#/50 m)	17.6	1.0	5.4	16.0	19.0
Stroke efficency (SEI units)	3.8	0.2	5.9	3.4	4.1

Figure 1. Timecourse of progression in competitive performance in the 100 m breaststroke.

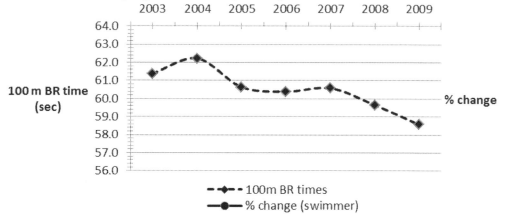

Interpretation of changes in performance, anthropometric and physiological test scores was undertaken using a quantitative approach accounting for the magnitude of the change, the typical error of measurement, and reference or threshold value for the smallest (worthwhile) change score (16). This approach provides the opportunity for interpreting changes in dependent measures in a rigorous quantitative manner rather than relying solely on subjective interpretations. In practice, the preferred approach is a combination of the coach's experience and the sports scientist's technical and analytical skills.

Conclusions

The 2009 World Champion in the Men's 100 m Breaststroke, Brenton Rickard, improved his level of competitive performance markedly in the two years prior to his success at age 25 y. The athlete possessed a mature physique by age 20 y and maintained good shape from season to season. A high level of consistency in physiological and performance measures obtained from the 5 x 200 m step test was evident between seasons, coupled with a cyclical pattern of improvement within a season leading to the major competition. The combination of kinematic, physiological, and biomechanical predictors of performance highlights the importance of an interdisciplinary approach to assessing training adaptations and performance. These patterns of change and variability in performance and test measures provide benchmark values for the long-term preparation of other breaststroke swimmers.

References

1. Pyne DB, Mujika I, Reilly T. Peaking for optimal performance: Research limitations and future directions. *J Sports Sci.* 2009,**27**:195-202.
2. Bosquet L, Montpetit J, Arvisais D, Mujika I. Effects of tapering on performance: a meta-analysis. *Med Sci Sports Exerc.* 2007,**39**:1358-1365.
3. Pyne DB, Trewin CB, Hopkins WG. Progression and variability of competitive performance of Olympic swimmers. *J Sports Sci.* 2004,**22**:613-620.
4. Stewart AM, Hopkins WG. Consistency of swimming performance within and between competitions. *Med Sci Sports Exerc.* 2000,**32**:997-1001.
5. Trewin CB, Hopkins WG, Pyne DB. Relationship between world-ranking and Olympic performance of swimmers. *J Sports Sci.* 2004,**22**:334-345.
6. Hawes MR, Sovak D. Morphological prototypes, assessment and change in elite athletes. *J Sports Sci.* 1994,**12**:234-242.
7. Pyne DB, Anderson ME, Hopkins WG. Monitoring changes in lean mass of elite male and female swimmers. *Int J Sports Physiol Perform.* 2006,**1**:14-26.
8. Anderson ME, Hopkins WG, Roberts AD, Pyne DB. Monitoring seasonal and long-term changes in test performance in elite swimmers. *Eur J Sport Sci.* 2006,**6**:145-154.
9. Pyne DB, Lee H, Swanwick KM. Monitoring the lactate threshold in world-ranked swimmers. *Med Sci Sports Exerc.* 2001,**33**:291-297.
10. Anderson ME, Hopkins W, Roberts A, Pyne D. Ability of test measures to predict competitive performance in elite swimmers. *J Sports Sci.* 2008,**26**:123-130.
11. Thompson K, Garland S, Lothian F. Assessment of an international breaststroke swimmer using the 7 x 200-m step test. *Int J Sports Physiol Perform.* 2006,**1**:172-175.
12. Thompson KG, Halljand R, MacLaren DP. An analysis of selected kinematic variables in national and elite male and female 100-m and 200-m breaststroke swimmers. *J Sports Sci.* 2000,**18**:421-431.
13. Seifert L, Leblanc H, Chollet D, Delignières D. Inter-limb coordination in swimming: effect of speed and skill level. *Hum Mov Sci.* 2010,**29**:103-113.
14. Leblanc H, Seifert L, Chollet D. Arm-leg coordination in recreational and competitive breaststroke swimmers. *J Sci Med Sport.* 2009,**12**:352-356.
15. Fritzdorf SG, Hibbs A, V.Kleshnev. Analysis of speed, stroke rate, and stroke distance for world-class breaststroke swimming. *J Sports Sci.* 2009,**37**:373-378.
16. Hopkins WG. How to interpret changes in an athletic performance test. *Sportscience.* 2004,**8**:1-7.

Commentary: *Kevin G. Thompson*, University of Canberra, Australia

The case study highlights how it is important to determine changes within kinematic parameters as well as physiological ones during competitions and step tests. It has been established from competition analyses that better 100-m breaststroke swimmers demonstrate greater competency in kinematic and temporal race components such as mid-pool swimming velocity, start times to 15-m and turning times (1). When the finishing times of 36 male international 200-m breaststroke swimmers differed by approximately 1.9% over two competitions it was observed that changes in their mid-pool swimming velocity, turning times and start time accounted for 60%, 34% and 6% of the difference; clearly demonstrating that improving each element is important in terms of improving overall race performance (2).

During a 100-m breaststroke race, the mid-pool swimming velocity tends to decrease significantly (~6-7%) as the race progresses demonstrating that the swimmer slows down over the course of the race due to fatigue (1). Male 100-m breaststroke swimmers tend to increase their stroke rate (by 3-4%) from the first to the second length to compensate for a shortening stroke length, however the reduction in stroke length is disproportionate (9-10%) and so the swimming speed worsens as the race progresses (1). Breaststroke swimming is mechanically inefficient and so increasing the stroke rate to compensate for a reducing stroke length will lead to an increasingly rapid onset of fatigue which is particularly noticeable in the leg muscles. Turning times will subsequently worsen due to a reduction in approach speed and a loss of motor skill (technical ability) and leg power. The case-study presented outlines how Brenton Rickard's stroke kinematics (stroke rate and stroke count) were monitored over a number of seasons to determine changes in stroke efficiency. This is good practice as faster race performances have been linked to the swimmer achieving higher mean stroke rates and greater mean stroke lengths (2, 3), however as stroke kinematics are unique to the individual they should not be directly compared between swimmers as this would be misleading (1).

The case study detailed how improvements in swimming speeds associated with various blood lactate concentrations (3mM - 8mM) were observed over the course of each season. Improvements in swimming speeds associated with blood lactate concentrations (from 2mM - 6mM) during 5 x 200-m step tests have been previously shown to coincide with improvements in 100-m and 200-m breaststroke race performances at national championships occurring 7-10 days later (N=11 swimmers), while the swimming speed associated with the peak heart rate attained during the final repetition of the step test did not change (3). Interpretation of step test data should be based on a quantitative approach in tandem with an in-depth knowledge of the swimmer's health and training status. Speed-lactate data have been observed to fluctuate following periods of successful training, overreaching, detraining and also poor nutritional practice (4). A race readiness test designed to assess absolute performance in temporal components (start time, turning time and overall time over 50-m) as well as speed endurance and turning consistency (2 x 100-m maximal efforts, 10 minutes apart) undertaken within 48 hours of a step test can further inform step test interpretation (5).

References

1. Thompson KG, Haljand R. An analysis of selected kinematic variables in national-elite male and female 100 m and 200 m breaststroke swimmers. *J of Sports Sci* 2000;**18**:421-431.
2. Thompson KG, Haljand R, Lindley M. A comparison of selected kinematic variables between races in national to elite male 200 m breaststroke swimmers. *J. of Swim Research* 2004;**16**;6-10.
3. Thompson KG, Cooper S. Breaststroke performance, selected physiological variables and stroke rate. *J. of Human Movement Studies* 2002;**44**(1):001-017.
4. Thompson KG, Garland S, Lothian F. () Interpretations from the physiological monitoring of an international swimmer. *International J. of Spos Science and Coaching* 2006; **2**(1):117-124.
5. Thompson KG, Garland S. Assessment of an international breaststroke swimmer using a race readiness test. *International J. of Spo Physio and Performance* 2009;**4**(1):139-143.

CHAPTER 27
Improving Performance of Arm Amputee Paralympic Swimmers

Carl J. Payton
Department of Exercise & Sport Science, Manchester Metropolitan University, Crewe, UK.

Vignette

In the International Paralympic Committee (IPC) Functional Classification System, swimmers with physical impairments are assigned to one of ten classes. The S1 class contains those with the most severe impairments; S10 contains those who are least physically impaired. Swimmers with a single arm amputation at elbow level are considered to have a relatively low level of impairment and therefore compete in the S9 class. A partial or total arm amputation reduces the surface area of the upper extremity and, consequently, the capacity of the swimmer to generate propulsion (1). In able-bodied front crawl swimming, there is evidence to indicate that the upper arm makes very little contribution to propulsion during the arm stroke (2). Despite this, all unilateral arm amputees pull their stump (upper arm) through the water in an attempt to gain propulsion. From observation of trained arm amputee swimmers, it is apparent that there is considerable inter-swimmer variability in the way in which the arm pull is executed; specifically, arm amputee swimmers vary in how rapidly they rotate the affected limb and in how they co-ordinate the pull of the affected limb with that of the sound arm.

As lead biomechanist to the Great Britain Paralympic swimming team, one of the author's key roles is to undertake applied research in order to provide coaches with objective evidence on which to base their coaching practice. There are currently seven unilateral arm amputees on British Swimming's World Class Programme; their coaches are constantly seeking to advance these athletes by improving the amount of propulsion they can generate. This case study will review the biomechanics research literature relevant to arm amputee swimming and conclude by highlighting the application of the research findings to coaching practice.

Discussion

Biomechanics research into front crawl swimming technique is quite extensive. However, the vast majority of studies have focussed on able-bodied swimmers and their findings therefore have limited practical application for the coach of an arm amputee swimmer. This short review will focus primarily on the few recent studies that have involved arm amputee swimmers.

Measuring propulsive forces in swimming is not a straightforward task. As these forces cannot be measured directly during unconstrained swimming, they must either be measured under conditions in which the swimmer interacts with some form of force transducer (3, 4, 5), or estimated using numerical techniques such as Computational Fluid Dynamics (6, 7, 8, 9, 10).

Payton and Wilcox (1) provided some evidence that arm amputees are able to generate propulsion with their upper arm. Eight trained unilateral arm amputee swimmers swam front crawl at distance pace (1.09 $\pm$ 0.13 m·s^{-1}) with a buoyancy aid between the legs in order to isolate the arm action. Trials were filmed underwater and intra-cyclic swimming speed (Figure 1) was measured in real time using a velocity meter.

Peak swimming speed during the push phase of the unaffected limb (1.30 $\pm$ 0.17 m·s^{-1}) was higher (*p*<0.05) than during the push phase of the affected limb (1.14 $\pm$ 0.11 m·s^{-1}), indicating that the swimmers were able to use their affected limb to increase their swimming speed within the stroke cycle, but not as effectively as with their unaffected limb. Interestingly, the speed at which the affected limb was pulled through the water appeared to be independent of the swimmers' stroke rate (*r*=-0.36). The authors observed that inter-arm coordination varied considerably between the participants, but did not attempt to quantify this variable.

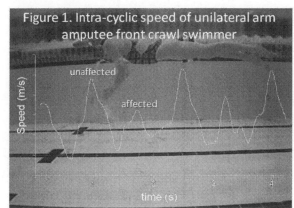

Figure 1. Intra-cyclic speed of unilateral arm amputee front crawl swimmer

Payton *et al.* (11) have demonstrated that unilateral amputee swimmers have greater shoulder extension strength on their unaffected side, compared to their affected side, although there is considerable variation in strength asymmetry between swimmers. The maximum shoulder extension torque generated by the affected and unaffected arms of nine well-trained female amputee swimmers was measured on an isokinetic dynamometer at $60°·s^{-1}$ and $180°·s^{-1}$. No association was found between the shoulder strength or the strength asymmetry of the swimmers, and their swimming performance level. This indicated that shoulder extension strength *per se* was not an important determinant of swimming performance in this group. With the same group of swimmers, Lee *et al.* (3) used a tethered swimming protocol to measure the propulsive force generated in a maximal effort (static) 30 second swim and compared the results to those from a matched group of able-bodied swimmers. Participants were attached to a load cell mounted on the end of the pool via a lightweight pole and a waist belt. The amputee swimmers produced approximately 20% lower mean tether forces than the able-bodied swimmers. However, the peak tether force produced by the unaffected arm of the amputees did not differ significantly from that of the dominant arm of the able-bodied swimmers. The tether force of the amputee swimmers declined during the test at a similar rate to that of able-bodied swimmers but their strokes rates declined at a significantly greater rate. The maximum propulsive force generated during the test (66.1 ± 3.2 N) correlated significantly with the amputees' 100 m front crawl best time ($r=-0.71$, $p<0.05$). A similar relationship has been reported for able-bodied swimmers (4, 5).

Tethered swimming provides a more ecologically valid measure of the arm amputees' force production capability than lab-based strength testing. Unfortunately, it is difficult to quantify accurately the contributions to propulsion made by each arm, during tethered swimming, due to the potential overlap of the propulsive phases of the two arms and the propulsion from the leg kick. Additionally, propulsive forces in tethered swimming can greatly exceed those acting during unconstrained swimming due to different fluid flow characteristics around the arm.

In the last ten years, Computational Fluid Dynamics (CFD) has emerged as a viable method of quantifying the contribution made by body segments/limbs to propulsion in swimming. CFD is a branch of fluid mechanics that involves solving fluid flow problems using computer simulations. Most CFD studies of arm propulsion have only considered the hand and forearm segments. Bixler and Riewald (6) found that the lift and drag coefficients produced by a hand and forearm CFD model, under steady flow conditions, compared well to coefficients determined experimentally in a water flume (12). Later studies have employed *unsteady* CFD analysis in which the effects of the acceleration of the hand (10) or hand and forearm (9) on propulsion are accounted for.

Two recent studies (7, 8) have used unsteady CFD specifically to evaluate the contribution made by the upper arm to the propulsion of swimmers with an amputation at elbow level. A CFD mesh model was produced (Figure 2) from laser scanning of a female arm amputee swimmer. Simulations were run to determine the effect of body roll, shoulder extension speed and swimming speed on propulsion. In most simulations, the upper arm generated propulsion. A typical shoulder extension speed generated a mean propulsive force of 3.5 N when swimming at 1.06 $m·s^{-1}$. This contrasts with mean propulsive hand forces of 8.1 ±1.4 N reported for able-bodied female swimmers (13). The model showed that propulsion was improved by an increase in body roll and an increase in shoulder extension speed. However, as the mean swimming speed was increased, the upper arm became progressively less able to generate propulsion. In some simulations, run at moderate swimming speeds, the upper arm created a net (resistive) drag force, not propulsion, during the pull.

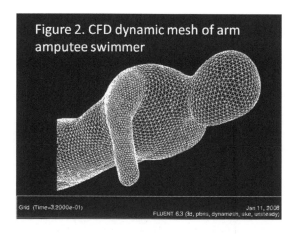

Figure 2. CFD dynamic mesh of arm amputee swimmer

In front crawl swimming, the correct coordination of the two arms is critical. Skilled able-bodied swimmers modify their inter-arm coordination according to their swimming speed. At slow speeds they adopt *catch-up* timing which involves a time delay between the propulsive phases of the two arms; at faster speeds they switch to *opposition* or *superposition* timing in which the propulsion from the two arms is continuous or overlaps (14).

Osborough *et al.* (15) examined the effect of swimming speed on inter-arm coordination in a group of thirteen well-trained unilateral arm amputee front crawl swimmers. The phase lag between the propulsive phases of the two arms was quantified using an Index of Coordination (IdC) adapted from the work of Chollet *et al.* (14). Five speeds ranging from 80-100% of each swimmer's maximum were used. An IdC < 0 denoted catch-up timing; an IdC > 0 represented superposition timing.

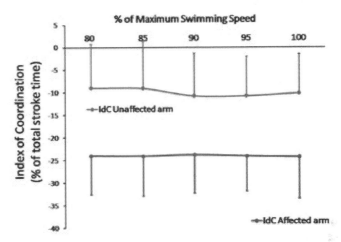

Figure 3. Index of Coordination (mean ± SD) for the affected and unaffected arms of 13 trained unilateral arm amputee swimmers at five relative swimming speeds

At all speeds, the amputees used catch-up arm coordination (Figure 3); this was greater on their affected-arm than their unaffected-arm, at all swimming speeds. This indicated a longer non-propulsive period following the pull of the unaffected arm, than following that of the affected arm. In contrast to able-bodied swimmers, the amputees' inter-arm coordination did not change significantly as swimming speed was increased. Interestingly, the fastest swimmers in the study exhibited greater inter-arm coordination symmetry (IdCs for affected and unaffected arms closer together) than the slower swimmers; this enabled them to achieve higher stroke rates. This is noteworthy since a previous study (16) demonstrated that stroke rate, rather than stroke length, is the most important determinant of performance for highly-trained arm amputee swimmers.

Conclusion

Recent biomechanical studies of arm amputee swimmers provide some important information for those who coach this group of swimmers and for those who design their strength and conditioning programmes. The key messages that have emerged are: 1) The upper arm can generate propulsion but, at fast swimming speeds, it may create resistance, 2) Propulsion from the upper arm is limited by shoulder extension speed; body roll can enhance propulsion from the upper arm, 3) Shoulder extension strength of amputee swimmers far exceeds that required to pull the affected arm through the water and so is not a limiting factor to performance, 4) Emphasis should be placed on developing high stroke rates. This can be achieved by reducing the non-propulsive period before the start of the affected arm pull. These guidelines have been disseminated to coaches and support staff through British Disability Swimming's Delivering on Deck Workshop programme. Making the required changes to the key biomechanical parameters, e.g. shoulder extension speed, stroke rate, body roll and IdC, is an ongoing process that is underpinned by the biomechanics support programme (17).

References

1. Payton CJ, Wilcox C. Intra-cyclic speed fluctuations of uni-lateral arm amputee front crawl swimmer. *Portuguese J of Sport Sci* 2006;**6** (Supl 2): 73-75.
2. Hay JG, Thayer AM. Flow visualisation of competitive swimming techniques: the tufts method. *J of Biomech* 1989;**22**:11–19.
3. Lee CJ, Sanders RH, Payton CJ. Changes in force production and stroke parameters of trained able-bodied and unilateral arm amputee female swimmers during a 30 second tethered front crawl swim.*J of Sports Sci* 2014; doi:10.1080/02640414.2014.915420.
4. Sidney M Pelayo P Robert A. Tethered forces in crawl stroke and their relationship to anthropometrics characteristics and sprint swimming performance. *J of Human Movement Studies* 1996;**31**:1-12.

5. Morouço P, Keskinen KL, Vilas-Boas JP, Fernandes RJ. Relationship between tethered forces and the four swimming techniques performance. *J of Applied Biomech* 2011;**27**(2):161-169.

6. Bixler B, Riewald S. Analysis of a swimmer's hand and arm in steady flow conditions using computational fluid dynamics. *J of Biomech* 2001;**35**:713-717.

7. Lecrivain GM, Slaouti A, Payton CJ, Kennedy I. Using reverse engineering and computational fluid dynamics to investigate a lower arm amputee swimmer's performance. *J of Biomech* 2008;**41**: 2855-2859.

8. Lecrivain GM, Payton CJ, Slaouti A, Kennedy I. Effect of body roll amplitude and arm rotation speed on propulsion of arm amputee swimmers. *J of Biomech* 2010;**43**:1111-1117.

9. Rouboa A, Silva A, Leal L, Rocha J, Alves F. The effect of swimmer's hand/forearm acceleration on propulsive forces using computational fluid dynamics. *J of Biomech* 2006;**39**:1239–1248.

10. Sato Y, Hino T. Estimation of thrust of swimmer's hand using CFD. In *Proceedings of the 8th Symposium on Nonlinear and Free-Surface Flows* 2002 pp71-75 Hiroshima.

11. Payton CJ, Naemi R, Machtsiras G, Sanders R. Isokinetic shoulder extension strength of trained female uni-lateral arm amputee swimmers. Presentation to the *16th FINA Sports Medicine Congress* Manchester, 2008.

12. Berger MA, de Groot G, Hollander AP. Hydrodynamic drag and lift forces on human/arm models. *J of Biomech* 1995;**2**:125–133.

13. Gourgoulis V, Aggeloussis N, Vezos N, Kasimatis P, Antoniou P, Mavromatis G. Estimation of forces and propelling efficiency during front crawl swimming with hand paddles. *J of Biomech* 2008;**41**:208-215.

14. Chollet D, Chalies S, Chatard JC. A new index of coordination for the crawl: Description and usefulness. *Int J of Sports Med* 2000;**21**:54-59.

15. Osborough C, Payton CJ, Daly D. Influence of swimming speed on inter-arm coordination in competitive unilateral arm amputee front crawl swimmers. *J of Human Move Sci* 2010;**29**:921-931.

16. Osborough C, Payton CJ, Daly D. Relationships between the front crawl stroke parameters of competitive unilateral arm amputee swimmers with selected anthropometric characteristics. *J of Applied Biomech* 2009;**25**(4):304-312.

17. Payton CJ. Biomech Support for British World Class Disability Swimming. *SportEX Med* 2008;**36**:9-13.

Commentary: *Nicholas Diaper, The English Institute of Sport, Loughborough, UK.*

Determining the limitations to performance in any sport and discipline is fundamental to the process of optimising performance. In the Paralympic context this is perhaps of even more importance given the fact that performance can be impaired by mechanical, physical or neurological restrictions that may not be encountered in able-bodied sport. In addition to this, practitioners and coaches working with Paralympic athletes may not necessarily have a Paralympic background. Therefore the challenge for all concerned, lies in understanding the limitations associated with a particular disability and the relationship with performance. Furthermore, not only do the physical and technical capabilities of the athlete need to be considered, but also the interaction between the athlete and the constraints placed upon them by the environment; in this case, overcoming the drag forces associated with movement through water.

Unfortunately for the Paralympic practitioner or coach, applied Paralympic-specific published evidence is few and far between. The tendency is often to look towards the able-bodied equivalent and apply the knowledge and findings to the Paralympic model. In some cases this approach *can* inform practice, but this particular case study clearly highlights the limitations of inferring able-bodied findings to Paralympic sport. Whilst some of the fundamental principles might be the same, there are clear differences that have significant implications for practitioners as well as coaches. Without this knowledge, improving performance is but a guessing game.

This work illustrates the value of applied scientific research in the field of Paralympic sport as it clearly contributes to our understanding of the limitations in optimising propulsion in unilateral arm-amputee swimmers. In turn, this knowledge provides critical evidence on which coaches and practitioners can base their interventions. It is also a reminder of the importance of the integration of sports science and sports medicine service provision, as perhaps the most explicit recommendation in this case is that developing high stroke rates is essential for optimising propulsion. The implications of this message are of great significance to other disciplines such as strength and conditioning, physiology, physiotherapy and of course coaching.

The role that technology, techniques and ideas from other industries can play in aiding our understanding of the limitations to Paralympic performance is also clearly demonstrated here. Although this may seem a stark difference from traditional Paralympic sports science disciplines, it is important for technology and new ways of thinking to be embraced. Improving the understanding of the impact of a particular disability on performance in this way, may also provide directions and considerations for the future development of functional classification procedures in Paralympic sport.

There is one final and important comment worthy of note despite the subtlety of it; one could argue that increasing our understanding of the limitations that disabilities have on performance, may also contribute to our understanding of able-bodied performance itself. As Paralympic athletes continue to push the boundaries of performance and challenge perceptions, perhaps it is time for able-bodied sport to look for answers from the Paralympic world?

CHAPTER 28
Competitive Versatility and Longevity of a World-class Triathlete

Iñigo Mujika, I.[1] and David B. Pyne[2]

[1]Department of Physiology, Faculty of Medicine and Odontology, University of the Basque Country, Leioa, BASQUE COUNTRY
[2]Sports Science and Sports Medicine, Australian Institute of Sport, Canberra, AUSTRALIA.

Vignette

A 36-year old male triathlete, Eneko Llanos from Vitoria-Gasteiz, Basque Country, has been competing and obtaining world-class results in Olympic Distance, XTerra and Ironman triathlon events since 1995. In this case study, we report on the long-term performances of this athlete. We gathered the athlete's competition results over the 18-y period between 1995 and 2012 inclusive, computed mean time, standard deviation and coefficient of variation (CV) for each event, and the swim, bike and run segments. Correlations between total time and each leg's time were also computed. Between 1995 and 2004, Llanos competed in 56 international Olympic Distance events (Olympic Games, World Championships, European Championships, World Cups and PanAmerican Cup). His average placing in these events was 17th, his total race time 113.2 ± 5.8 min (mean ± SD), CV = 5.1%. The correlation between total race time and the swim, bike and run legs was r = 0.20, r = 0.78 and r = 0.44 respectively. Between 2000 and 2012 Llanos competed in 20 international XTerra events, including 13 World Championships. His average placing in these events was 6th (including 3 World Champion titles, 2 runner-up positions and two XTerra U.S.A. victories): his total race time was 155.5 ± 8.4 min, CV = 5.4%, with correlations of swim (r = 0.02), bike (r = 0.87), and run (r = 0.85). In 2005 Eneko started his Ironman triathlon career, taking part in 16 events over 7 years (of which 5 World Championships), placing 4th on average (four wins and five runner-up placings), a total race time of 497.7 ± 15.8 min and CV = 3.2% The correlations between total time and swim, bike and run times were r =0.52, r = 0.92, and r =0.82 respectively. The competitive results over different distances and race formats in a career stretching for 18 years is indicative of the athlete's outstanding competitive versatility and longevity. The bike and run legs were the most important in achieving the highest levels of international performance in various triathlon events.

Discussion

Triathlon is a multisport event consisting of swimming, cycling and running. International triathlon competition is primarily divided in three events:
1. Olympic Distance races, comprising a 1.5-km open water swim, a 40-km bike leg allowing participants to ride in group, and a 10-km road run;
2. XTerra events, raced over identical or similar distances, but characterised by a mountain bike ride and a cross-country run;
3. Ironman races, consisting of a 3.8-km open water swim, a 180-km cycling ride in which drafting behind other athletes is not allowed, and a 42.2-km run.

Top level male athletes typically cover Olympic Distance events in ~110 min, XTerra events in ~150 min, and Ironman events in ~500 min. Elite competitors usually specialise in one of these events (11), although a limited group of athletes have shown their prowess in more than one triathlon format.

The subject of this case study is a highly versatile world-class triathlete, able to excel in Olympic Distance, XTerra and Ironman events, obtaining major victories at the world stage in the three specialities. Llanos was the International Triathlon Union Long Distance World Champion in 2003, in his first ever long distance triathlon race and at a time when he was otherwise exclusively racing in the Olympic Distance. He was also able to win the 2007 European Triathlon Union Cross Triathlon European Championships when he was exclusively dedicated to the Ironman distance. Additionally, he has won what is known as the Hawaiian Double (fastest combined time at Ironman World Championhips in Kona and XTerra World Championships in Maui) consecutively from 2006 to 2010. Llanos also won the 2010 Inaugural Abu Dhabi Long Distance triathlon, raced over a novel distance of 3.0 km swim, 200 km bike and 20 km run, providing further evidence of his ability to adapt to different racing formats and distances.

This versatility may seem somewhat counterintuitive, considering that the physiological demands of the various triathlon formats are substantially different. In the shorter, draft-legal events, athletes typically swim very fast to make the front pack in the bike leg (13, 16). The bike leg is characterised by a highly variable power output profile of a somewhat stochastic nature, interspersing short bouts of supramaximal power output, with longer periods of sustained maximal and submaximal power output (3). Long distance events, on the other hand, require a more moderate power output under steady state conditions. Nevertheless, international standard short distance and long distance specialists exhibit similar maximal and submaximal physiological attributes during cycling and running (11), making the competitive versatility of Eneko Llanos more understandable. In addition, Llanos was exposed to a variety of sports and outdoor activities since his early childhood (e.g. judo, hiking, rock climbing, skiing, various ball games), and this variety has continued over the years, particularly in the off- and early-season. Participation in a range of sports of an aerobic nature may provide the physiological stimuli required for gross adaptations that could be easily transferred to triathlon (1).

The reliability of competitive performance in sports characterised by stable environmental conditions such as swimming, running, and weightlifting is high in top level athletes, with CV values ranging from 1.2 to 3.1% (6, 9, 14, 15). When conditions are variable, however, performance variability is much larger (10). The CV of Llanos' overall performance (5.1% in the Olympic Distance, 5.4% in XTerra and 3.2% in Ironman races) is relatively small, considering that triathlon competition takes place over highly variable race courses and environmental conditions (Table 1). The CV for different legs of different events ranged between 3.3 and 12.3%, and Llanos appears to be quite consistent on the bike leg of all events (known to be one of his strengths as a triathlete). In contrast Llanos was least consistent on the swim leg of the Olympic Distance events, often a limiting factor for competition placing (13, 18). In contrast, his consistency in the marathon run of Ironman races has been a major determinant for his success in this distance.

Table 1. Descriptive analysis of the competitive history of World-Class triathlete Eneko Llanos (Basque Country) over a 18 y period in three different triathlon formats. SD, standard deviation; CV, coefficient of variation.

Event				Total	Swim	Bike	Run
Olympic Distance	Average Placing	17	Mean Time (min)	113.2	18.9	60.3	33.1
	Number of Races	56	SD	5.8	2.3	5.2	3.0
			CV (%)	5.1	12.3	8.6	9.0
			Correlation		0.20	0.78	0.44
XTerra	Average Placing	6	Mean Time (min)	155.5	20.1	92.6	42.3
	Number of Races	20	SD	8.4	1.7	5.1	4.1
			CV (%)	5.4	8.3	5.5	9.7
			Correlation		0.02	0.87	0.85
Ironman	Average Placing	4	Mean Time (min)	497.7	48.8	273.9	171.1
	Number of races	16	SD	15.8	2.8	10.1	5.7
			CV (%)	3.2	5.7	3.7	3.3
			Correlation		0.52	0.92	0.82

In agreement with a previous investigation (4), Llanos' swim performance showed the lowest correlation with the overall finish time, particularly in shorter Olympic Distance and XTerra events (Table 1). In the Ironman, however, the swim leg correlated quite well with finish time (Figure 1), while the bike leg showed the largest correlation with finish time for all three events, probably because it makes up a bigger proportion of total race time.

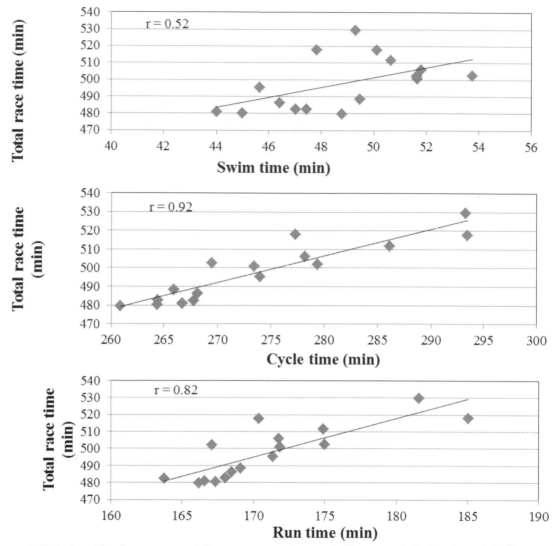

Figure 1. Relationships between total Ironman race time and: Top panel) Swim time; Middle panel) Bike time; Bottom panel) Run time.

One of the factors arguably contributing to Llanos' competitive longevity is his resilience to injury. Up to 60% of elite Olympic Distance and 75% of Ironman athletes have been reported to be affected by injury, and 43% of the British Senior Squad suffered traumatic injury over a five-year follow-up (17). In contrast, Eneko's only major injury keeping him from training in the past 18 years was shin splints back in 2005, when he switched the focus of his training from the Olympic to the Ironman distance. In addition, his excellent bike handling skills have probably protected him against traumatic injury over the years.

Another key factor is the training and competition planning. High-level long distance triathletes' training is characterised by a systematic design and structure, both in the short- and long-terms. This pattern allows expert athletes to perform greater amounts of training, maximise physiological adaptations and avoid overtraining and burnout (1). A periodised approach to training and competition has been a priority for Eneko and his coach (I. Mujika) over the years, with a particular emphasis on training load-recovery interaction (7, 8), tapering in the lead-up to major events (12), and recovery after each macrocycle and each competitive season (5). Llanos' proactive cognition (i.e. the ability to identify opportunities and act on them to bring about meaningful change), is a predisposing personality characteristic eliciting higher levels of performance (2).

Conclusions

World Champion triathlete Eneko Llanos of the Basque Country has been successfully competing at the world stage in the Olympic Distance, XTerra, Ironman and other triathlon events for 18 y with outstanding versatility and longevity. His versatility can probably be attributed to an early exposure to different sports and outdoor activities, which has continued over the years, and long-term adaptation to multiple stimuli of an aerobic nature. Llanos' consistency of performance (low variability) within and between events is a key feature of his career. Moderate to high correlations between total race times and each of the three legs of a triathlon indicate that Eneko is a very capable cyclist and runner, although the swim appeared to be a limiting factor in his Olympic Distance career. His longevity is most likely the result of his resilience to overuse injury, excellent bike handling skills that have limited traumatic injury, effective training and competition planning, and proactive cognitive skills during racing.

References

1. Baker J, Côté J and Deakin J. Expertise in ultra-endurance triathletes. Early sport involvement, training structure, and theory of deliberate practice. *J Appl Sport Psychol.* 2005,**17**:64-78.

2. Baker J, Côté J and Deakin J. Cognitive characteristics of expert, middle of the pack, and back of the pack ultra-endurance triathletes. *Psychol Sport Exerc.* 2005,**6**:551-558.

3. Bentley DJ, Millet GP, Vleck VE and McNaughton LR. Specific aspects of contemporary triathlon: implications for physiological analysis and performance. *Sports Med.* 2002,32:345-359.

4. Dengel DR, Flynn MG, Costill DL and Kirwan JP. Determinants of success during triathlon competition. *Res Q Exerc Sport.* 1989,**60**:234-238.

5. Gould D, and Dieffenbach K. Overtraining, underrecovery, and burnout in sport. In M. Kellmann (Ed.), *Enhancing recovery: Preventing underperformance in athletes* 2002 (pp. 25–35). Champaign, IL: Human Kinetics.

6. Hopkins WG, Hewson DJ. Variability of competitive performance of distance runners. *Med Sci Sports Exerc.* 2001,**33**:1588-1592.

7. Issurin V. Block periodization versus traditional training theory: a review. *J Sports Med Phys Fitness.* 2008,**48**:65-75.

8. Issurin VB. New horizons for the methodology and physiology of training periodization. *Sports Med.* 2010,40:189-206.

9. McGuigan MR and Kane MK. Reliability of performance of elite Olympic weightlifters. *J Strength Cond Res.* 2004,**18**:650-653.

10. Mendez-Villanueva A, Mujika I and Bishop D. Variability of competitive performance assessment of elite surfboard riders. *J Strength Cond Res.* 2010,**24**:135-139.

11. Millet GP, Dréano P and Bentley DJ. Physiological characteristics of elite short- and long-distance triathletes. *Eur J Appl Physiol.* 2003,**88**:427-430.

12. Mujika I and Padilla S. Scientific bases for pre-competition tapering strategies. *Med Sci Sports Exerc.* 2003,**35**:1182-1187.

13. Peeling P and Landers G. Swimming intensity during triathlon: A review of current research and strategies to enhance race performance. *J Sports Sci.* 2009,**27**:1079-1085.

14. Pyne DB, Trewin CB and Hopkins WG. Progression and variability of competitive performance of Olympic swimmers. *J Sports Sci.* 2004,**22**:613-620.

15. Stewart AM and Hopkins WG. Consistency of swimming performance within and between competitions. *Med Sci Sports Exerc.* 2000,**32**:997-1001.

16. Vleck VE, Bentley DJ, Millet GP and Bürgi A. Pacing during an elite Olympic distance triathlon: comparison between male and female competitors. *J Sci Med Sport.* 2008,**11**:424-432.

17. Vleck VE, Bentley DJ, Millet GP and Cochrane T. Triathlon event distance specialization: training and injury effects. *J Strength Cond Res.* 2010,**24**:30-36.

18. Vleck VE, Bürgi A and Bentley DJ. The consequences of swim, cycle, and run performance on overall result in elite Olympic distance triathlon. *Int J Sports Med.* 2006,**27**:43-48.

Commentary: *Charles R. Pedlar,* St Mary's University College, Twickenham, UK

Mujika and Pyne present a rare and fascinating case report of an exceptional athlete, demonstrating a clear case of an athlete who has achieved a long and successful career with exceptional consistency and dedication. Such high average placings in highly competitive events (17th, 6th and 4th over Olympic

Distance, XTerra and Ironman events respectively) are not down to luck or talent alone, rather, this level of consistency demands an immensely professional and rigorous approach of which other athletes and support staff should take note. The most successful endurance athletes coming through our lab present with similar characteristics. These characteristics include:

1.   a progressive training programme;
2.   balancing overload and recovery in a measured systematic fashion; .
3.   a desire to apply new techniques and gain advantages no matter how small;
4.   a resilience to injury, possibly related to a variety of multi-directional, aerobic based sports undertaken in the early years.

Llanos was able to perform over a relatively short distance (winning the 2007 Cross Triathlon European Championships) whilst training for an Ironman event, which raises the question of the value of high volume training for relatively short events. Although the specific physiological determinants of performance continue to be examined and discussed (1), typically, high volume training over a number of training years improves exercise economy whilst maximal aerobic capacity remains high but constant, as demonstrated in a case study of Paula Radcliffe (2). In the present case study of Eneko Llanos, performance in shorter distances may well have been influenced by the improvement in exercise economy gained from training for the longer event. We have demonstrated that running economy is an important factor even in middle distance running events (3), and many other case studies exist where a high volume of training over many years leads to improvements even over shorter distances. Paula Radcliffe achieved her best results over 3, 5 and 10 km distances on the track and the road within 1 year of her marathon world record.

Llanos has already achieved a competitive career of 19 years. We continue to see improvements in endurance physiology well into an athlete's 4th or even 5th decade, provided the motivation is strong. There are some famous examples. Mike Dixon, a British Olympian achieved world class performances at 6 consecutive winter Olympics (1984 – 2002) in the biathlon event and continues to train and compete. Steve Redgrave achieved Olympic Gold in rowing at a record breaking 5 consecutive Summer Olympic Games (1984 – 2000) and many athletes go on to compete in senior competition across the lifespan (4). There is no reason why Llanos should stop here!!!!

References

1. Reilly, T, Morris, T, Whyte, G. The specificity of training prescription and physiological assessment: A review. *J of Spo Sci* 2009; **27**(6): 575–589.
2. Jones, A.M. The physiology of the world record holder for the Women's Marathon. *I J of Sports Sci and Coach* 2006;**1**(2): 101-114.
3. Ingham, S. G, Whyte, G. P, Pedlar, C, Bailey, D. M., Dunman, N, Nevill, A. M. Determinants of 800m and 1500m running performance using allometric models. *Med and Sci in Sport and Exe* 2008;**40**(2):345-350.
4. Wright, V.J, Perricelli, B.C. Age-Related rates of decline in performance among elite senior athletes. *Ame J of Sports Med* 2008;**36**: 443.

CHAPTER 29

The Erotic Adventures of D: Interactions with a Triathlete

David Tod[1] & Mark B. Andersen[2]

[1] School of Social Sciences, University of the Sunshine Coast, Queensland, Australia
[2] School of Sport and Exercise Science and the Institute for Sport, Exercise and Active Living, Victoria University, Melbourne, Australia

Vignette

When Kate (a pseudonym), a 19-year-old triathlete, walked into my office I was struck by her beauty. If asked to describe my ideal female, I would have pointed to Kate. I was a 22-year-old neophyte in the first few months of my career as a sport psychology practitioner, and Kate had come to find out how I could help her prepare for her first international race in 10 months time. I was delighted to be recognised as having expertise and to be working with a talented athlete. The first session went well; I was guided by Taylor and Schneider's (1) intake interview guide, and over the 90 minutes together we discussed how mental skills could help her prepare for the event, and we explored various topics. We arranged to meet again the following week.

As Kate sat down at the start of the second session, she dropped a letter on my desk and asked me to read it. The letter said that she was grateful for the previous week's session and that I was the first person who had shown her true love. She had not ever been loved before, even by her parents and family, and she hoped that I would continue to love and care for her because she was a fragile and sensitive person.

My immediate reaction included anxiety and the realisation that I was out of my depth. I explained to Kate that as a sport psychology consultant I dealt with performance and not personal issues (a naive and false dichotomy). I knew a counsellor, however, who did deal with personal issues like the ones about which she had written. I asked Kate if she would be happy to meet the counsellor, and after she said yes, we walked together to the counsellor's office and made an appointment.

I continued to work with Kate for the following months until her competition. Our sessions were typically long, and I looked forward to them because I felt we had clicked and had built a strong rapport. Kate reported making good progress in counselling and after a number of months developed a romantic relationship with a male athlete whom I was also helping.

Discussion

The decision to refer Kate was a proper one, because my training had equipped me to educate clients about mental skills that could assist them with their sports performances. I had not been trained for much else. I was self-aware enough to recognise the boundaries of my competence and not let my erotic countertransference lead me to attempt to treat Kate's feelings of being unloved (and maybe unlovable).

The decision also forced me to deal, as best I could, with my feelings of guilt and shame. I had become overwhelmed with the fantasies I had experienced. Throughout the previous week I had pictured Kate naked, had wondered what she was like to kiss, and had fantasised about romantic intimate contact with her. My mental movies had been followed by shame and guilt. I understood that sexual liaisons are harmful in the therapeutic context, are an abuse of power, and are destructive for clients' (and my) welfare (2). I also felt I had violated the Christian principles my parents had taught me (more shame).

My reflections were narrow and misinformed by a focus on sex. Although the erotic is typically equated with sexual attraction, it involves much more than physical desire. An erotic experience may involve multiple desires at play (e.g., closeness, interconnectedness; 3). Consciously, I recognised that I was erotically attracted to Kate. It was not until I reflected on the experience after termination of the consulting relationship, that I came to understand that I was also attracted to Kate because she helped me feel competent. Kate worked hard at her mental preparation, applied the strategies we discussed during her physical training and competitions, and gave me positive verbal feedback about my assistance. I interpreted these signs as indicators that I was good at my job.

My focus on my own reactions (anxiety, excitement, arousal, shame), coupled with my narrow view of applied sport psychology, interfered with my level of empathy for Kate. My surprise at the way Kate reacted after the first session was a sign of my limited appreciation of the potentially deep interpersonal connections that may develop in service to others. With the benefit of experience, I can better understand Kate's reaction to our initial meeting and appreciate how she might have viewed me as a stand-in romantic (or even parental) love object for what had been missing in her life (4). I had displayed a caring non-judgemental attitude, a focus on her welfare and interests, and an absence of a self-serving agenda. The letter she wrote probably indicated that my caring, unconditional approach had been a profound experience for Kate, and she likely felt relieved to learn it was possible, acceptable, and safe for her to share personal sensitive material with another person; love was not confined to romantic movies and books. And how did I respond? Metaphorically, I kicked her when she was down, and communicated that some material is not acceptable and should remain unmentioned. That Kate focused on mental skills training after session two and that we largely did not delve into her love issues (attributable, in part, to the help she was receiving from my colleague) was probably influenced by the way I responded to her when she did share her sadness and longing.

After the session when Kate dropped her letter on my desk, I contacted a mentor for advice. Early in my career, formal training and supervision pathways had not been established in the country where I was working at the time). Individuals with minimal training were helping athletes and, in many cases, were receiving no supervision. I was fortunate to have a mentor who helped me when needed. My mentor operated from a psycho-educational perspective in which practitioners assisted athletes with their performance, but not their personal, issues. He agreed with my decision to refer Kate to a counsellor and reinforced my justification that it was an ethically sound course of action. My mentor, however, did not help me deal with my fantasies and emotions, or reflect on how I may have influenced the situation. I also did not admit to these feelings and thoughts, because I worried he might have thought I was a suspect practitioner for having them. It was not until I read some psychodynamic literature and received mentoring from a psychodynamic-oriented practitioner (second author) that I worked through my own countertransferential erotic material and gained greater insight into my interactions with Kate. With this help, I was able to accept that my fantasies were normal human reactions, that I was not a pervert or an unethical practitioner, and also that I could manage my thoughts and emotions so that I could stay focused on helping clients.

Although I largely enjoyed working with Kate after session two, I also felt some anxiety when meeting her because I worried I would reveal, unintentionally or unconsciously, my erotic feelings. Not sharing my erotic countertransference with Kate was a correct decision, because it would have placed her in an unreasonable position; dealing with my thoughts and emotions was my responsibility, not Kate's (4). Given that I had no experience in working through my longings and emotions, and no supervisor who might have been equipped to help me, suppression and avoidance were my only coping strategies. By engaging in self-surveillance and self-monitoring, I may have been less authentic and *present* with Kate than I might have been otherwise, and this may have limited our working alliance to some degree.

Through my experience with Kate, and several of my other early clients, I realised that to be effective, I needed to be able to do more than just teach clients to use mental interventions (e.g., goal setting, imagery, self-talk, relaxation) to enhance performance if I were going to help them with their issues (5). My realisation led me to return to formal education and was also a reason for the topic I chose when completing a PhD (i.e., how sport psychology graduate students develop into competent, effective, and ethical practitioners).

Conclusions

Athletes and sport scientists (not just sport psychologists) will likely experience erotic desires and longings when interacting with each other. Clients are often young, athletic, attractive, and outgoing. Many sport science consultants are caring, non-judgemental individuals interested in their clients, and that care (a form of love) can be attractive to athletes (6). Across the sport sciences, erotic material in applied service is probably seldom discussed in training or supervision beyond prohibitions against sexual misconduct. Such taboos and avoidance of discussing the erotic in athlete-consultant relationships may leave practitioners fearful of voicing their desires with mentors and unable to manage their erotic thoughts and feelings, as well as, potentially, their emotions of anxiety, guilt, shame, and confusion. Open discussion about such material, however, may help prepare practitioners for a phenomenon that is likely to occur sooner or later (7).

References

1. Taylor J, Schneider BA. The sport-clinical intake protocol: A comprehensive interviewing instrument for applied sport psychology. *Prof Psychol Res Practice* 1992:**23**:318-325.
2. Yalom ID. *The gift of therapy: Reflections on being a therapist.* London: Piatkus; 2001.
3. Stevens LM, Andersen MB. Transference and countertransference in sport psychology service delivery: Part I. A review of erotic attraction. *J Appl Sport Psychol* 2007;**19**:253-269.
4. Mann D. *Psychotherapy, an erotic relationship: Transference and countertransference passions.* New York: Routledge; 1997.
5. Tod D, Lavallee D. Taming the Wild West: Training and supervision in applied sport psychology. In: Gilbourne D, Andersen MB, eds. *Critical essays in applied sport psychology.* Champaign, IL: Human Kinetics; 2011:193-215.
6. Andersen MB. Touching taboos: Sex and the sport psychologist. In: Andersen MB, ed. *Sport psychology in practice.* Champaign, IL: Human Kinetics; 2005:171-191.
7. Tod D, Andersen MB. Practitioner-client relationships. In: Mellalieu SD, Hanton S, eds. *Professional practice issues in sport psychology: Critical reviews.* London: Routledge; 2012: 273-306.

Commentary: *Richard Godfrey,* Brunel University, UK

As sports scientists from any discipline, we can from time to time be subject to temptation. Accordingly, I empathise with the experiences, challenges and conflicts indicated in Tod and Andersen's case study. I too have been confronted by situations where there is a risk of desire rendering me unprofessional. On a number of occasions, during the 12 years I worked for the British Olympic Association, I was propositioned by female athletes. Although it is incredibly flattering to be thought of as attractive by another attractive human being, from a professional perspective caution is required. Often the athletes concerned are vulnerable, either as a result of their ages or their career desires (they want to be part of the team and represent their country) and the perception that you represent authority. This latter point is particularly important as power and authority are themselves powerful aphrodisiacs and, as sports scientists, we must remind ourselves of that fact. Perhaps it is that power and authority, rather than us as individuals, that athletes are attracted to.

Most sports scientists seek careers in this area stemming from their own enthusiasm for sport. For the younger sports scientist, when working as part of a team of support staff, they very often align with the athletes they might themselves once have had aspirations to be and to whom they are, generally, closer in age than some of their support staff colleagues. We are there, however, because of our jobs, and so, generally, it is best to avoid potentially compromising situations and remember to maintain professional distance from athletes who, in reality, are akin to *clients*.

That mutual attraction should develop is not surprising and, in most cases, between adults in this environment, it is neither wrong nor unhealthy. Nevertheless, even where consent is forthcoming, in my opinion, in this context it is not correct to act upon one's feelings. If the attraction is so strong that a romantic relationship, and all it entails, is unavoidable then a change of circumstance is required. That is, resigning from work with that particular sport or team demonstrates integrity and arguably eliminates professional concerns (including conflict of interest). Professional misconduct is a serious accusation and one which could not only result in job loss but, the reputation which will subsequently follow you around, could make getting another job difficult, or worse, find yourself facing disciplinary proceedings. I suggest practitioners examine the professional code of conduct on this issue, but the take home message is that prevention of a relationship starting is better than trying to find a way to make it work. However, if mutual attraction is strong then sacrificing work in that environment is advisable. It demonstrates commitment both to the other person and to one's career.

Any relationship that might develop with an athlete should be considered as serious; serious, because of the professional issues involved. It has always been my contention that all support staff, whatever their parent disciplines, should attend, at least, the most basic of counselling courses. This could provide support staff with greater skills with which to do the right thing in many sensitive situations. After all, we are all human and subject to the same desires.

10877063R00096

Printed in Great Britain
by Amazon.co.uk, Ltd.,
Marston Gate.